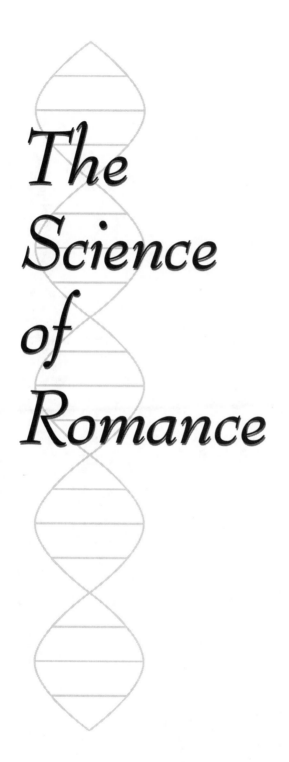

The
Science
of
Romance

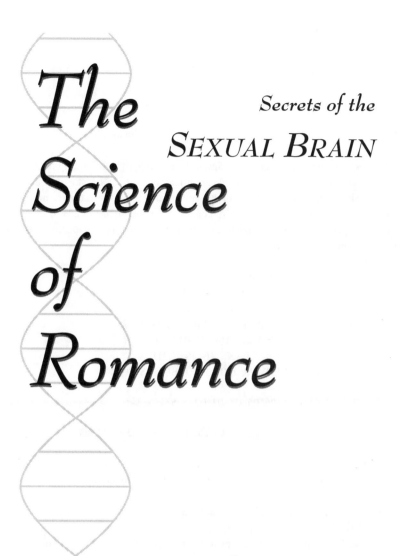

The Science of Romance

Secrets of the
SEXUAL BRAIN

Nigel Barber

 **Prometheus Books**

59 John Glenn Drive
Amherst, New York 14228-2197

Published 2002 by Prometheus Books

Inquiries should be addressed to
Prometheus Books
59 John Glenn Drive
Amherst, New York 14228–2197
VOICE: 716–691–0133, ext. 207
FAX: 716–564–2711
WWW.PROMETHEUSBOOKS.COM

06 05 04 03 02   5 4 3 2 1

Library of Congress Cataloging-in-Publication Data

Barber, Nigel.
    The science of romance : secrets of the sexual brain / Nigel Barber.
        p. cm.
    Includes bibliographical references and index.
    ISBN 1–57392–970–0 (alk. paper)
    1. Sex. 2. Sex (Psychology) 3. Sex differences (Psychology) 4. Man-woman rela-
tionships. 5. Genetic psychology. I. Title.

HQ21 .B184 2002
306.7—dc21

2002021367

Printed in Canada on acid-free paper

Contents

Acknowledgments

I am thankful for the help received from many people at various stages in the conception, planning, and writing of this book. My literary agent, Elizabeth Frost-Knappman, has lived and suffered through various stages of this work during the proposal phase. I am grateful to her for her unwavering faith in the commercial viability of this project and to her partner at New England Publishing Associates, Edward W. Knappman, and their assistant Kristine A. Schiavi.

My research has been facilitated by the work of the Interlibrary Loan Department at the Portland Public Library. Loan staffers Rita Gorham, Eileen MacAdam, and Anne Ball were unfailingly cheerful and helpful. I appreciate their professional dedication. I was also helped by Maja Keech of the Library of Congress Prints and Photographic Division, who assisted with copyright issues.

I am grateful to my editor at Prometheus Books, Linda Regan, for many thoughtful comments on the manuscript, and to Linda Mealey, Lynn Pullano, and Mary A. Read, although any errors must remain my responsibility. Trudy Callaghan has also provided editorial assistance and has been quick to identify any inadvertent gender biases. I appreciate her continued advice and support.

I am grateful to many friends and colleagues in research whose breathtaking findings in the scientific study of human sexuality during the past few decades have put evolutionary psychology on the intellectual map. Rather than naming all the members of the Human Behavior and Evolution Society whose research and ideas have influenced my own work, I will name just one, Donald Symons, whose courageous 1979 book, *The Evolution of Human Sexuality*, has been and continues to be an inspiration to all of us. It makes an overwhelming case for the value of a scientific approach to romance.

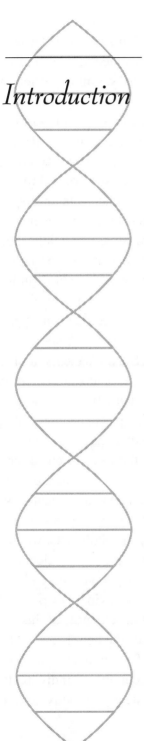

Introduction

The scientific approach to the study of sexual behavior and psychology begins with Charles Darwin, but it certainly does not end there. Darwin was primarily concerned with the question of human origins, but he was also a pioneer in evolutionary psychology. He recognized that human displays of emotion, for example, were connected to the expression of emotion in other animals and that the human body has been modified by sexual selection for the purpose of attracting mates.

It was very difficult for Darwin to accept these ideas himself. He is famous for his scholarly reserve, withholding publication of his most important ideas on the theory of evolution by natural selection for some twenty years, until he was spurred into action by a nasty surprise on June 18, 1858. On this date, he received in the mail a letter from a young naturalist, Alfred Russell Wallace. His letter not only contained a condensed version of Darwin's theory, in general terms, but even used strikingly similar terminology. Among other shared influences, it turned out that both men had read Thomas Malthus's work that described the inevitability of starvation in a world where population always grew at a more rapid rate than the available food supply. From there, it was a small step to speculate that certain kinds of animals or people would be better able to survive and reproduce than others, and would transmit their beneficial traits to offspring.

Darwin's hesitation was more than academic and scientific. It is clear that he had trouble accepting a view of human beings that was at odds with the one accepted by his forebears for centuries. Metaphorically speaking, he spent twenty years pinching himself to be sure that he was awake.

Since that time, evidence from the fossil record, from molecular biology, from physiology, and from evolutionary psychology has accumulated, and such fits comfortably within Darwin's theory. It provides a sketchy but multifaceted picture of organic evolution in general and of human evolution in particular. We are gradually becoming acquainted with ourselves as an evolved species on this planet and the realization is often a little shocking—rather like the experience of reading a scandalous novel and discovering that one of the naughty characters was obviously based on ourselves. There is a feeling of shock, of recognition, and at the same time, a reluctance to admit that it is really us.

Some of the most compelling recent evidence on human evolution has been assembled by evolutionary psychologists, whose discoveries have blown a lot of cobwebs from the social sciences—for example, the implausible view that humans are infinitely elastic, or "plastic," and that the variation in our behavior and psychology from one society to another is arbitrary. One example of this fallacy is the tired idea that beauty is in the eye of the beholder. A substantial body of evidence in the new field of Darwinian esthetics confirms that there is overwhelming agreement among individuals and societies about what makes people physically attractive, although there are also some differences in tastes from one individual to another or from one society to another. The many traits of physical appearance that make people attractive converge on a few simple biological ideas, such as health and fertility. Physically attractive people of the opposite sex can be defined as those with whom we would want to invest our reproductive efforts.

Many of the discoveries in Darwinian esthetics have been widely disseminated in the popular press because they are both fascinating and accessible. It is no surprise to discover that symmetrical people, whose left and right sides of the body and face match up with each other, are more physically attractive than those who are "lopsided," or physically disproportionate. But the underlying mechanism according to which symmetrical individuals combine good genes with a favorable early environment is particularly intriguing. Add to that the discovery that symmetrical men produce a chemical attractant that may make them sexually irresistible to women, and we find ourselves in a twilight zone wherein the real world seems stranger than science fiction.

Part of the fascination of any scientific pursuit involves recognition of the *strange* in the midst of the *familiar*. Evolutionary psychology deals with the very familiar realm of ordinary experience, but makes this realm strange by showing that it is organized around themes of competition for survival and reproduction—of which most of us are generally unaware. In that sense, it is like Alice's looking glass that allows us to enter the strange yet recognizable world of evolutionary explanations. Like Wonderland, evolutionary psychology has its own compelling logic, but it is subtly different from the logic of everyday life; that is why it can seem so odd.

Whether one looks at recent discoveries that pertain to human reproduction—such as physical attractiveness, dating competition, causes of divorce, sexual jealousy, sexual differences in responses to erotica, brain chemicals mediating romantic love, or airborne chemicals known as *pheromones*—there is a dazzling array of new findings and ideas relevant to the romantic evolutionary past of our species.

Evolutionary social scientists have spent a lot of effort attempting to convince their colleagues that natural selection has designed people to respond in predictable ways to similar situations. This now seems so obvious that it may hardly seem worth mentioning. To take a simple example, the expression of facial emotion is pretty much the same in every society: the same basic emotions are expressed using the same pattern of contractions of facial muscles elicited by the same type of situation. When people encounter an old friend, they are happy and smile a lot, whether they belong to the Fore tribe of New Guinea, or are urban residents of Japan. Charles Darwin suspected that this was the case and confirmed his idea by asking correspondents around the world to describe how local people expressed their emotions. In recent times, emotion researcher Paul Ekman and his colleagues have conducted more objective tests with the same results.

Modern evolutionary thinking goes beyond Darwin in suggesting that natural selection shapes not just our predictable species-inherent qualities, but also our ability to respond to change. This point is well illustrated in two aspects of change that affect our love lives. One is change during the development of the individual. The other is change in romantic attitudes and behavior over the course of history, or from one society to another.

The first kind of change is illustrated by the fact that children who grow up in a socially stressful environment are thereby programmed for vulnerability to various social problems, such as teen motherhood, low occupational achievement, and criminal activity. Long viewed by psychologists as an instance of individuals being crushed by a difficult early environment—such as being raised in extreme poverty—these phenomena were reinterpreted by Jay Belsky and his colleagues at the University of Pennsylvania as an example of people developing in ways that help them confront the challenges of survival and reproduction under difficult social circumstances. When the going gets tough, the tough get going. Children learn to live in the present, taking risks and grasping at pleasure, and they learn to look out for number one. Crime is thus an extreme example of putting one's own interests before those of victims and the community. Teenage childbearing is also an adaptive response to a scarcity of marriageable men, which is associated with poverty and unemployment. Given that marriage to men who can provide reliable economic support for a family is unlikely, young women might as well begin their families while they are young, healthy, and energetic. That is why teen motherhood is so common in economically depressed regions.

A stressful childhood changes romantic relationships. They become more intense, more theatrical, and more unstable. It is probably no accident that many leading entertainers, from Marilyn Monroe to Billie Holiday, who were notoriously unhappy in their romantic lives, also had difficult childhoods.

The second kind of change to which evolutionary thinking brings order and coherence is societal change. Recent history has seen astonishingly rapid changes in virtually all aspects of marriage and sexual relationships. Many of these changes can be interpreted as the increasing economic power of women altering the balance point of their dependence on masculine support. This has greatly weakened the stability of the marriage bond in different ways, including the fact that fewer children are raised. (According to worldwide data, couples with four or more children are ten times less likely to divorce than couples with none.)

The rising economic status of women, and the sexual liberation that goes along with it, might seem to put modern societies beyond the limits of evolutionary explanation—which, after all, describes our adaptation to two million years as hunter-gatherers. But further examination reveals that modern urban women are in some ways closer to hunter-gatherers than to agriculturalists. For example, women provide close to two-thirds of the food consumed in a hunter-gatherer family, and the vegetable food they gather thus dwarfs the caloric value of meat hunted by men. This means that their economic power is arguably greater than that of men. Women in these societies enjoy considerable sexual freedoms compared to agricultural communities, and extramarital affairs are common. Interestingly, successful hunters are popular as extramarital sex partners, suggesting that these affairs have economic as well as sexual motives because good hunters have more meat to distribute and likely favor their lovers. Marriages are also comparatively unstable. So, from the point of view of relations between the sexes, urban families of the twenty-first century have a great deal in common with our subsistence-level ancestors.

Be that as it may, the really exciting aspect of evolutionary social science is that it helps us to account for differences *between societies* in the same terms as historical change in a *particular* society. A convenient illustration is provided by the changing standards of bodily attractiveness for women. A plumper ideal body type is admired in societies where women have little social power and are valued mainly for their fertility. In societies where women are economically important because their work is highly valued, or because they own property, a more slender figure is considered attractive. The point is that standards of attractiveness vary in a way that helps people to survive and reproduce in their specific social context. Exactly the same logic can be applied to explaining the dangerously extreme standard of slenderness adhered to in our modern society compared to standards common in the agricultural past, when women had little economic power.

The underlying logic for the changing ideal of women's bodily attractiveness in different societies and at different historical periods is that it is based on evolved responses to body shape. Thus, women having highly feminine, highly curvaceous bodies are seen as sexually attractive and highly capable of reproduction (for which there is good medical support) but as lacking in professional competence (for which there is absolutely no evidence). In fact, the bias against curvaceous women's perceived professional competence is far stronger than the bias against overweight persons, judging from experimental data. It is small wonder that writers of "success" manuals for women should advise them to dress in tailored business suits that de-emphasize the secondary sexual traits of their bodies. Adherence to a less curvaceous ideal is also strongly correlated with entry of women in the professions, where they are expected to play down their sexual signals to increase perceptions of career competence.

Men also mute their secondary sexual traits, most drastically by shaving. Why they do so is less clear, although preliminary research has linked changes in marriage to changes in facial hair fashions. At times when women have difficulty marrying and when single parenthood rates are high, men are most likely to shave their mustaches. This pattern probably has little to do with sexual attractiveness, as such, but to women preferring clean-shaven men, it is a way of selecting men who are "open books"—emotionally transparent, reliable, and faithful. This would eliminate the faithless cads who are just out for sexual enjoyment.

The relative difficulty that women experience in marrying has at times been a factor in their entry into professions and can even account for some aspects of change in dress fashions. During the 1950s, economic prospects in the United States brightened and women turned away from jobs, settling down to marry and raise children. The fundamental factor behind this shift was the sudden availability of a large number of single men who were economically qualified to support a family. Simultaneously, the fashion of bodily attractiveness for women reversed trend from slenderness to the moderately voluptuous ideal represented by actresses of the period such as Jane Russell. The hemlines of skirts also lengthened as ordinary women wanted to convey a message of sexual reserve that appeals to potential husbands, but not to men who are interested in casual relationships.

The marriage market changed dramatically during the 1960s. With a scarcity of marriageable men, we got sexual liberation, short skirts, feminism, the entry of large numbers of women into professions, and the reemergence of the slender standard of attractiveness. The connection between each of these phenomena can be explained in terms of a relatively simple theory that relates evolved bodily signals and evolved relations between men and women to changes in the marriage prospects of each sex. The great advantage of assessing societal change in the light of evolutionary adaptations is that we can detect patterns that were previously unnoticed. If something is orderly, we can begin to explain it in scientific terms. Conversely, if we do not detect such patterns, we face the dismal prospect of concluding that changes in social trends for sexual behavior and romantic relationships are arbitrary and incomprehensible. Although the romantic behavior of our species may have subjective elements that elude science, it also shows unmistakable patterns that bear the imprint of their evolutionary origins. Such patterns make up the substance of this book.

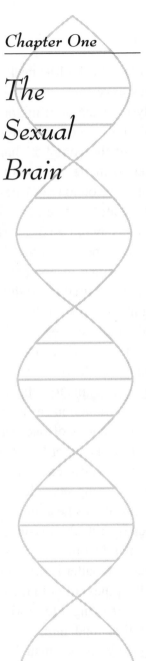

Chapter One

The Sexual Brain

Romance is biology. It serves the biology of reproduction. It is etched into the biology of the brain. Differences between men and women reveal the brain's basic relation to romance in interests, aptitudes, and desires. Such differences loom large because they confront us every day, appear in all societies, and are astonishingly resistant to change.

In a *Seinfeld* episode entitled "The Voice," Jerry, portraying a crazed lover overcome by romantic love, runs among the pigeons that are courting furiously even as they scatter before him. The cooing of pigeons is a cinematic cliché used to summon up associations of young love in springtime. In reality, pigeons are indeed a romantic lot, falling in love and sticking together through thick and thin.

Take, for example, the ringdove, a close relative of the domestic pigeon, which has a distinctive black ring around the back of its neck. Students of courtship behavior in ringdoves have found that the persistent cooing of the male is not nearly as inane as it might first seem. In fact, the strutting of the male and his cooing call as he chases a female is an obligatory feature of the complex chain of events in the reproduction of this species. If the female is not vigorously courted by the male, she will not move on to the successive phases of reproduction: selection of a nest site, nest building, copulation, laying eggs, and incubation. After a day or two of courting, the female shows signs of interest toward the persistent male. She flips her wings in a distinctive way and begins to approach the male. Careful research has shown that male courtship produces a chemical response that can be measured in the blood of the female. Once the nest site has been selected, the male occupies it and produces a different call, known as the nest coo. When the female begins emitting her own nest coo, showing that she is ready to start building the nest, the male stops cooing and begins to collect nest material that the female then arranges. Reproduction in the ringdove is like a complex dance in which the actions of one partner affect the biology, and hence the behavior, of the other.[1]

To emphasize the biological basis of sex differences in psychology and behavior is not the same as endorsing sex-role stereotypes any more than an anthropologist who studies head hunting is inviting people to go out and shrink heads. Yet, the existence of sex-typed adaptations in the brains of men and women implies that the ideal of a society in which gender does not matter is (a) not possible and (b) most likely contrary to the happiness of both men and women. We know this because of the harrowing life stories of people whose sex was reassigned in early life, condemning them to live the life of a woman while possessing the brain of a man. Exclusive homosexuality, which evidently has a biological basis for women as well as men, provides further evidence that sex roles are partially present in the brain at birth.

SEX DIFFERENCES IN THE BRAIN

Sex differences in the brain programmed men and women to behave in ways that contributed to survival and reproduction in the evolutionary past. Men have skills that would have helped them to be effective hunters, for example, and women have superior manual dexterity, which would have helped them to be effective gatherers. Yet it is important to realize that behavioral differences based on sex are not entirely built into the brain at birth for human beings any more than they are for ringdoves. Sex-typed behavior is affected by experience as well as by biological sex. Many psychologists refer to such experiential effects on sex-typed behavior as "gender" but the term is often used misleadingly because of the assumption that what we learn is distinct from our biology. In reality, all learning is associated with changes in brain biology, and what we are capable of learning and motivated to learn is affected by the evolution of the brain.[2]

The relationship between brain and behavior is a two-way street. The brain controls behavior, but behavior may also change the brain. For example, experience in playing sports could develop three-dimensional abilities and thereby help children to excel in some branches of mathematics, such as geometry. There is even evidence that girls who have more experience in team sports have superior mathematical abilities. We do not know, however, whether there is a cause-and-effect relationship here. It might be that young women with exceptionally good spatial abilities would be drawn to playing sports like basketball, baseball, and soccer. Yet, there is no evidence that men who are unusually good at sports are also good at mathematics.

Nature and nurture are often not the antagonistic forces that psychologists of the past believed. People often gravitate toward doing things that they are good at and thereby develop those talents.

Recent findings in behavioral neuroscience and in psychology clearly demonstrate that the brains of men and women process information differently. These phenomena are often controversial,

but we can no longer dismiss them as a reflection of sex-role pre-
conceptions of scientists. Consider the following:[3]

- The left cerebral cortex, which houses language, is thicker in
  women while the right cerebral cortex, housing spatial ability,
  is thicker in men.
- Women recover language function better after left-hemi-
  sphere damage than men do.
- Men have faster simple reaction times than women; for
  example, they are faster to push a single button after a light
  flashes. The differences in reaction time are statistically reliable and range from about 10 to 20 percent, depending on the
  specific test procedure. However, in tasks involving recognition of complex patterns, women are faster.
- Men are much better than women at mental rotation tasks,
  such as imagining what an object looks like from above when
  it is shown only from the front.
- When quizzed about the location of objects in an office in
  which they had waited, women scored up to 70 percent better
  than men. In such a test of "incidental" memory, participants
  do not even know that they will be tested, and believe they
  are just waiting for the experiment to begin. According to
  Irwin Silverman and Marion Eals at York University in
  Ontario, this difference may represent an inherited adaptation
  of women to finding food by foraging. For example, if they
  remember the exact location of different types of fruit trees
  and the location of ripe fruits on the tree, they will be more
  efficient at harvesting them when needed.[4]
- Men are more likely to be left-handed, reflecting greater
  dominance of the right side of the masculine brain; this could
  be due to greater exposure of the male fetal brain to testosterone. Three-dimensional spatial abilities are also localized
  in the right side of the brain.
- PET scans (position emission tomography) are a method of

imaging energy use in the brain. Brain cells that are more active in processing information use more sugar as energy. PET scans can thus be used to determine which part of the brain is active during a particular mental activity. When asked to spell a word, men use the left hemisphere. Women use both sides of the brain.

- Consistent with greater use of different areas of the brain for a given task, women have thicker connections between the left and right sides of the brain (although this finding has been controversial in the past due to inaccuracies of measurement).
- Women feel sadness and some other emotions more keenly than men. For example, they suffer from depression about twice as often as men. Recent studies suggest that this difference is linked to more activity in structures of the brain collectively known as the limbic system, which regulates emotions.
- Women have better skills in nonverbal communication. They send more signals. They are better at reading the emotional content of a photograph or the emotion conveyed by tone of voice in experiments where the meaning of speech is garbled. They are also better at detecting deception through body language cues.
- Homosexual men have a smaller clump of cells (known as INAH3) in the hypothalamus than heterosexual men. The structure is closer in size to that of women. This suggests, but does not establish, that sexual orientation is a module determined by early brain development. If so, then the tendency of heterosexual women to be sexually attracted to men and the tendency of heterosexual men to be attracted to women may be attributable to anatomical sex differences in the brain that are present at birth.

Much of the foregoing evidence shows that brain differences are relevant to sex roles. For example, women's nurturant roles as caregivers and socializers of children are favored by communicative skills and emotional sensitivity. Men's lower emotionality and better three-dimensional skills may favor hunting, aggressive competition with peers, and warfare.

To say that these differences exist is not to say that they cannot be changed or that they apply to the same extent for everyone. For example, there is a sex difference in spatial ability of rats and other rodents mirroring humans' spatial abilities. In rats, the sex difference is based on fewer neural inputs to the hippocampus (a structure of the brain) for females. Janice Juraska of the University of Illinois at Urbana-Champaign has found that when female rats receive toys to manipulate, the innervation (nerve input) of the hippocampus increases. A parallel human finding—just noted—is the recent discovery that girls who are active in team sports do better at math. According to this view, declining sex differences in math may represent changing lifestyles for young women.[5]

Much of the controversy about the biological basis of brain differences stems from thinking about the brain as though it were a computer. Some aspects of the computer model are appropriate. For example, it appears that sexual orientation for men is fixed early in life and changes little. In that sense, it might be appropriate to think of sexual orientation as analogous to a microchip that is etched in early development as a result of the brain's exposure to sex hormones. Many aspects of brain function, however, are constantly being modified by environmental experiences from outside and by hormonal influences from inside. Some of the more dramatic examples of adult brain plasticity include the following:

- When children learn to play stringed instruments, the area of the brain cortex "answering" to the fingers of the left hand gets larger, according to the work of Ed Taub and others at the University of Alabama, Birmingham. The left hand is used to finger the strings producing different notes. The right hand, which clasps the bow, maintains a constant grip so that the fingers have less to "learn."[6]
- The architecture and connections made by individual cells in the brain are constantly changing. This allows for some recovery of function in stroke victims.

- Hormones produce constant change in neurons. For example, Estrogen promotes sprouting of new connections (dendrites). In addition, men do better on mental rotation tasks in the spring, when testosterone levels are lower, than in the fall, when they are higher; this seems perverse when you consider that male superiority in mentally rotating objects is due to the early influence of testosterone on the fetal brain.[7]
- Men are much more likely to commit crimes of violence at ages when their testosterone level is highest (late teens to early twenties); this shows that the hormone primes the human brain for aggression.
- Testosterone level is also a primary determinant of sex drive in both sexes. As men age, their testosterone level declines and sex drive is somewhat reduced. As women approach middle age, their testosterone level rises relative to estrogen and libido tends to increase.

The generally higher level of sexual behavior seen in men compared to women is clearly relevant to their greater proclivity for casual sex, but it might not be entirely due to different levels of testosterone in the blood. Other plausible alternatives include sex differences in brain anatomy and brain chemistry, parental inculcation of personal modesty in girls, and anatomy and physiology of the genitals.

ADAPTIVE SEX DIFFERENCES IN THE BRAIN

Just as women excel at remembering the location of objects, a skill that is directly relevant to foraging, men have better motor skills—at least those that are used for hunting. In all hunter-gatherer societies studied by anthropologists, men do most of the hunting of large animals, although women are often involved in collecting or trapping small animals, in fishing, and in driving game animals into traps. It is very unusual for women to be involved in killing large game animals by throwing projectiles at them such as spears or rocks, or striking them with precisely targeted weapons such as arrows from a bow or poisoned darts from a blowgun.[8]

Natural selection is the name of the process through which species acquire the bodily and behavioral characteristics that help them to survive and reproduce. Naturalists have always been impressed by the astonishing congruence between the bodily design of animals and the way this design enables them to live. The hummingbird's beak is exactly the right shape and length to reach the nectar at the base of a deep flower on which it feeds. The astonishingly long necks of giraffes similarly explain what separated the successful giraffe from less successful members of the species in reaching the leaves of tall trees. The key point about natural selection is that it is a means by which unsuitable types get weeded out. Giraffes with unusually short necks are not successful at getting enough to eat and are less successful at surviving and reproducing; their genes are therefore selected out of subsequent generations, making it unlikely that short-necked giraffes would be born. In a similar vein, men who were inept hunters had difficulty attracting a mate and left fewer children to perpetuate their incompetence in bringing in game. Being a poor hunter did not have such dire consequences for ancestral women, however.

If natural selection had acted on the brains of men to improve their success as hunters, then men should be considerably better than women at throwing objects precisely enought to hit a target. According to the research of Doreen Kimura at the University of Western Ontario, London, there is a large sex difference which favors men in tasks of precise targeting. On a dart-throwing task, for example, the average man did better than 84 percent of the women even though success at the task did not require strength.[9]

It would be hard to reduce this finding to any simple difference between men and women in muscle function, size, or strength. The results indicate that it was due to a genuine difference in brain processing of visual-spatial information. One might imagine that the difference could be explained in terms of the greater participation of men than women in sports like baseball and basketball, where targeting ability plays an important role. Yet, when Kimura used statistical control of sports participation level, she found that it accounted for only a trivial portion of the sex difference. Whatever their participation level in sports, men performed much better than women on tasks involving targeting.

Consistent with the argument that men have superior visual-spatial processing capacity, Kimura found that they were also better at blocking a Ping-Pong ball with their hands after it was flung from a launcher, although the sex difference in this task was only half as large as for the targeting activity. Speed of reaction might seem to offer an advantage here, but statistical control showed that the sex difference was not due to women having slower reaction times. Men excel in the higher level integration of visual and motor information, which makes sense in terms of behavioral adaptations because they would have used such aptitudes as hunters.

What makes these findings so fascinating is that women do not have poorer motor coordination than men. In fact, when you look at tasks requiring fine manual dexterity, women are reliably better than men. According to conventional tests of dexterity, this advantage is not simply due to women's greater manual adroitness because of the smaller size of their fingers. Kimura's research has shown that women perform much better in tasks that require them to control fingers individually such as bending a finger at the joint without moving any of the other fingers.

Again, this advantage of women in fine manual dexterity is consistent with the kind of work they perform most frequently in subsistence societies. These activities include gathering of small fruits and nuts, weaving baskets, making thread, and producing clothing. It is no accident, perhaps, that men are not often drawn to needlepoint and quilting as hobbies and are rarely seen in the local knitting circle. It is true that they might be made to feel uncomfortable if they cultivated such stereotypically feminine hobbies, yet this begs the question of why such hobbies are associated with women. Presumably, women cultivate them because they tend to be good at them, perhaps as a side effect of their evolutionary development of foraging skills.

Despite men's average deficiency in manual dexterity compared to women, many musicians are men. There is no question that playing a musical instrument is a severe test of manual dexterity. Male musicians may have unusually good manual dexterity to begin with, or they may develop it through practice.

It has been noted that—in addition to testing higher on targeting activity—men test better on mental rotation tasks than women. In mental rotation tests, people mentally rotate some object and decide what it would look like if viewed from another angle. Given that men evolved as hunters who traveled farther from the home base in search of game animals, it is possible that their superiority on spatial rotation tasks reflects an adaptation for finding their way around the landscape. Kimura and her colleagues tested this idea by asking subjects to learn a route through a streetmap. The experimenter first traced the route using a stylus that left no mark. Participants were then asked to trace the same route from memory. When they made a mistake, the experimenter corrected them and recorded the error. Men learned to trace the route correctly after fewer trials, and they made fewer errors before learning. Better scores on the map task were correlated with better spatial rotation scores, suggesting that the masculine advantage in some spatial aptitude tests may be due to a male adaptation for finding prey throughout their territory. Although it is approximately ten thousand years since most humans made their living by hunting, this is only one half of one percent of our two-million-year history as a species. Genetically speaking, we are virtually identical to our hunter-gatherer ancestors.

Not only did men and women in Kimura's study learn the map task at different rates, but after it had been mastered, they had absorbed different kinds of information. Men acquired more knowledge about distances and directions. Women recalled more details about landmarks and street names, reflecting their superior memory for the location of objects. This sex difference influences the way people give directions. Men operate as though referring to a mental map. Typical instructions are: "Go north three blocks, take a right on Union Avenue, go east four blocks, take another right on First Street, you will see it on the right half way up the block." Women's directions tend to be more concrete, vivid, and detailed. Instead of referring to a mental map, it is as though they were actually perceiving details on the route they are describing. When giving directions, a woman is more likely to refer to concrete objects along the route: "Go to the post office, turn right (that's Union Avenue); keep going until you see McDonald's, turn right on the wide street with trees on both sides—that's First Street. Ours is the white house with blue trim halfway down the block."

What makes the adaptationist (i.e., evolutionary) account of sex differences in spatial orientation so compelling is that a similar phenomenon has been observed in voles, or field rodents. Among species in which males range farther than females, they have an advantage in maze learning, reflecting superior spatial skills, but there is no sex difference for species having the same-sized home range. Among laboratory rats, males perform better on mazes than females and are more likely to rely on geometric cues, whereas females rely more on landmarks such as scratches on the walls of the maze.

The difference in rotational ability between sexes has been observed in all societies studied, with the interesting exception of the Inuit. The fact that women and men do not differ in spatial ability tests might have something to do with the exceptional visual challenge of living in a featureless tundra landscape that is constantly changing as snow drifts move about. Parents of both sexes must be vigilant in order for children to survive. Young children of the Inuit differ, however, showing that the difference is probably present at birth, reflecting early sexual differentiation of the brain. The demands of the environment evidently promote spatial ability of Inuit women until they equal those of men.

The effect of changing levels of hormones in the blood on sex-differentiated abilities is complex. When women are at the high-estrogen phase of their menstrual cycle, they perform better on tasks like verbal fluency and fine motor control, for which there is a feminine advantage. Men's testosterone level varies considerably over the course of a year, with a peak in the fall and a trough in spring. So much for springtime love and pigeons cooing. Men with generally high testosterone levels have poor spatial abilities. Men's scores on manual dexterity, verbal fluidity, and other female-favoring tasks do not change with the seasons. Taken together, these results provide compelling evidence that early brain differences are the underlying cause of differences in cognitive abilities between men and women. Apparently, sex hormones can modify these differences later in life also. These sex-related brain differences are of great interest in their own right, but they also provide compelling evidence that the brains of men and women have evolved differently. This helps to explain differences in sexual behavior between the sexes. As the French say, *viva la difference!*

CHANGING GENDER

If the thesis that romance is largely reducible to biology has any merit—for both human beings and ringdoves—then differences in the sexual behavior of men and women could have been modified by natural selection. Many important sex differences in the human brain are already present at birth; this has been demonstrated for animals such as monkeys. Artificially exposing female monkey fetuses to high levels of sex hormones in the womb masculinizes their brains and turns them into "tomboys," who are far more interested in rough and tumble play than normal females are. Such experiments could not be done on humans for ethical reasons (and some ethicists argue that they should not be performed on animals either). If we are to establish the important role played by sex hormones in producing differences in the romantic behavior of men and women, then we are forced to rely on natural experiments. In a natural experiment, researchers study the effects of an event that happened to occur, instead of looking at the results of a change they produced. Since natural experiments are less tidy than true experiments, scientists have less confidence in their conclusions and are forced to make a more convincing case by triangulating the results of several different studies.

The life stories of children who for various reasons have their sex misassigned (or reassigned) soon after birth highlight the importance of early sex differences in the brain. Genetic males are sometimes raised as females, thereby pitting their biological sex against the effects of rearing in a natural experiment. This might occur when the genitals are ambiguous in appearance.

One famous case involved male identical twins whose story first emerged in 1984. The penis of one of the twins was accidentally destroyed in a botched circumcision. Two psychologists at Johns Hopkins University, John Money and Anke Ehrhard, were called in to counsel the parents and they described the case in a scientific paper.[10]

Advised by doctors that there was no way to reconstruct the penis, the parents made a difficult decision to reassign the sex of the child. At the age of seventeen months, the boy was castrated and reconstructive surgery was carried out producing the outward semblance of female genitalia. He was given a feminine name, dressed in girls' clothing, and otherwise treated as a girl.

The mother was so pleased with the success of the operation that she could say, "She seems to be daintier. Maybe, it's because I encourage it. . . . I've never seen a little girl so neat and tidy. . . . She just loves to have her hair set."[11] Despite a positive evaluation of the effects of the operation by the parents, Money and Ehrhardt did see some problems. The girl was a tomboy. She had a lot of physical energy, was very active, could be very stubborn, and tended to dominate over female companions of the same age.

Despite these clear signs of problems for the reassignment, psychologists welcomed what they interpreted as a positive outcome because it supported their cherished belief that nurture can prevail over biology: Sex differences are just roles that men and women play like actors following a script. They have nothing to do with biology. They are really just "gender" differences.

Margaret Mead misguidedly promoted the belief that sex roles were learned rather than biological. She cited societies like the Tchambuli in New Guinea, in which men act like women and women behave like men. She claimed that Tchambuli men were submissive and anxious, while Tchambuli women were bold and domineering. Mead got it very wrong. In fact, Tchambuli men had multiple wives, whom they purchased through payment of a bride price. Husbands also ruled the domestic roost, feeling free to beat their wives whenever they felt dissatisfied. In all of the societies studied by anthropologists, men are more politically powerful and more physically aggressive than women. Women are more nurturant and spend more time and effort than men in caring for children and ministering to the needs of other adults. Rearing experiences may well modify these differences, but their universality has a very simple explanation. The brains of men and women are different at birth and this affects what they want to do.[12]

Psychologists have often misinterpreted the case history of the twins referred to earlier. The altered twin was not the happy success story that they, like the parents, wanted to see. In fact, he had never fit comfortably into the role of a girl. His unfeminine behavior had made him a laughingstock among peers in high school, who amused themselves by mimicking the lurching gait of the "cave woman." Devastated by his unenviable situation, he contemplated suicide at the age of fourteen. At this point, the parents finally admitted that he had been born a boy.

Following this revelation, he understood for the first time who and what he was. His life began to make sense. He decided to change back to being a boy again. This might have seemed like a heroic choice given that the change was to be carried out while he was still in high school, subject to the cruel jibes of peers. The transformation actually went quite smoothly because he found more acceptance as a boy than he had previously received as a girl. He made the calculation that things could not get any worse, and he was right.[13]

Males were also raised as females, as in the case of eighteen genetic males in the Dominican Republic who suffered from an unusual type of androgen insensitivity. At birth, these individuals seemed like females with slightly swollen clitorises so they were raised as girls. With the testosterone spurt at puberty, their clitoris-like sex organs grew and developed a penislike shape. Moreover, all of the normal secondary sexual characteristics of puberty, including muscular development of the torso, deepening of the voice, and the sprouting of facial hair occurred. These unusual people are called *Machihembra*, meaning (approximately) "first woman, then man."[14]

The Machihembra also raise the question of whether the biological sex of males can be neutralized by the experience of being raised as a female. The outcomes have generally been clear-cut. Most of the Machihembra have adopted masculine roles following puberty. Many have taken up stereotypically masculine occupations as miners and lumberjacks, although some people might interpret this as overcompensation for a perceived lack of masculinity. Some have even found female sexual partners and have been married. Here, biology obviously prevailed over rearing experiences.

Another unusual condition that reveals the effects of sex hormones on the brain and sex-role behavior is known as congenital adrenal hyperplasia (CAH, also known as adrenogenital syndrome). In this inherited condition, the part of the adrenal gland responsible for testosterone production grows too much, resulting in the production of excessive hormone. In males, this may result in early puberty but has no other obvious consequences. Females suffering from the condition are born with an enlarged clitoris and partly fused vaginal labia that may require corrective surgery. Once the condition is detected, the individual is treated with a synthetic hormone that suppresses testosterone production.

Treated CAH females are particularly interesting to developmentalists because they provide clues about what effects prenatal exposure to androgens (i.e., male sex hormones) can have on the brain; they therefore point to features of sex roles which might be biologically determined before birth. Girls with CAH turn out to be much more "tomboyish" than others. They are more likely to engage in rough-and-tumble play and are more active in sports like football that involve rough bodily contact.

Even toy preferences are affected. We now accept that boys and girls have clear toy preferences that emerge in spite of socialization pressures. Most boys prefer to play with vehicles and construction toys, and most girls prefer to play with dolls and stuffed animals. These preferences are not taught by parents, as many socialization theorists have claimed, and the toy preferences of CAH females highlight this conclusion. They show a distinct preference for "boy" toys over "girl" toys. It is hard to imagine that their parents would encourage this propensity. If anything, parents would resist these atypical toy preferences. Despite any interference by parents, CAH girls spend more time playing with vehicles and less time playing with dolls. They are even less interested in babies than girls who don't have this condition. In fact, they show no more interest in younger children than boys do.[15]

HOMOSEXUALITY AND SEX ROLES

CAH also has an impact on sexual orientation. Of thirty young women studied by psychologist John Money, eleven described themselves as homosexual or bisexual, twelve said they were exclusively heterosexual, and seven refused to talk about their sex lives. Excluding the uncommunicative women, almost half the CAH women (48 percent) were homosexual or bisexual. This is about five times the normal incidence of homosexual experience, according to the Kinsey report on female sexuality. If the women's failure to discuss their sex lives was associated with a problem of sexual orientation, the incidence of homosexual behavior could be as high as 60 percent.[16]

Exposure to unusually high levels of sex hormones thus has a masculinizing effect on the brains and behavior of CAH women. Based on their own reports and on the responses of their parents, CAH girls were more involved in athletic competition than normal young women. Approximately three-quarters of them were more active in sports than unaffected peers. This compares with a sex difference for unaffected individuals in which 90 percent of boys engage in more sports activity, on average, than girls, even though more females participate in sports today than in the past. CAH girls are not only masculinized in their behavior, they are also defeminized to some extent as reflected in their lack of interest in small children.[17]

CAH suggests that typical male behavior is produced by early exposure of the brain to testosterone. Yet, CAH is a genetic disorder and it is theoretically possible that the disorder, alters behavior through genetic mechanisms that have nothing to do with sex hormones.

For this reason, it is illuminating to compare the behavior of CAH females with that of others who have been prenatally exposed to high levels of sex hormones. Some pregnant women take the drug diethylstilbestrol (DES) during pregnancy to prevent miscarriage. DES is a synthetic form of estrogen. Although it is usually thought of as a "feminine" hormone, estrogen is not only present in males, but it is actually responsible for masculinizing their brains. The key to this paradox is that testosterone can be easily converted into estradiol (an estrogen), and it is estradiol that makes brains masculine. Early in development, a substance (alpha fetoprotein) protects females from the potentially masculinizing effects of their own estrogen. It binds to estrogen in the bloodstream, preventing it from entering the brain. Exposure to high levels of any sex hormone may overwhelm this defense mechanism, causing the brain to be masculinized.

DES can masculinize the brains of female fetuses that are genetically normal. That is, it can make them resemble the brains of men. One study of thirty women exposed to DES found that seven were homosexual or bisexual, compared to only one of a control group of thirty women.[18] This provides further evidence that sex hormones produce sexual attraction to women by early masculinization of the brain and thus illustrates why heterosexual men are sexually attracted to women.

The brains of homosexual men are in some minor ways similar to those of women, but they are generally more similar to those of heterosexual men. Although their brains are not fully "masculinized," they are not feminized either. They do not resemble the brains of women in most respects. Incomplete masculinization might happen if the male brain received less than the usual amount of stimulation by sex hormone at some point around the third month of gestation but was exposed to the usual amount at other times during development. Concrete evidence for the incomplete masculinization view comes as anatomical evidence showing that homosexual men are as similar to heterosexual women as to heterosexual men for specific sites in the brain. Thus, the anterior commissure, one of the connections between the two halves of the brain, is larger in women and in homosexual men than it is in heterosexual men. The suprachiasmatic nucleus of the hypothalamus is also larger in heterosexual men than it is in homosexual men or in women. This region houses the biological clock, and it is interesting that women tend to go to bed earlier and get up earlier than men, even when they have no children to care for—a difference that is also seen between heterosexual and homosexual men.[19]

These differences are interesting but of questionable relevance to sexual preferences because neither of these brain structures has been implicated in sexual orientation nor in behavior. That is why the finding of Simon LeVay—that a particular nucleus (clump of cells) in the hypothalamus known as INAH3 differs between homosexual and heterosexual men—is so important. This sexually dimorphic structure, normally smaller in women, regulates mammalian sexual behavior. Although there was much variability in his results, LeVay found that the INAH3 nucleus was twice as large in heterosexual men as in heterosexual women. The nucleus was almost the same size for homosexual men as for women.[20]

We do not know whether the nucleus studied by LeVay is responsible for sexual orientation in humans, but the work does suggest a mechanism through which male homosexuality could be produced. Suppose that there is a sexual-orientation nucleus in the brain, and that exposure to a high level of sex hormones early in development causes it to increase in size. As a result, the individual develops an attraction to the bodies of women. If the structure does not increase in size, the individual grows up with a sexual preference for men. Such a mechanism may not provide a perfect explanation for sexual orientation. Thus, three of the nineteen heterosexual men studied by LeVay had large INAH3s and three of the sixteen heterosexual men had small ones. Yet, it is a good beginning.[21]

What makes the incomplete masculinization interpretation of male homosexuality so compelling is that there are a number of superficially unrelated phenomena that fit in with this hypothesis. One is the circadian rhythm effect, the daily, twenty-four-hour activity cycle of many organisms. Another is that the advantage enjoyed by men in target skills is not found among male homosexuals according to the work of Doreen Kimura and her colleagues. In tests, they hung a vertical target on the wall and asked participants to strike it by pitching a Velcro-covered ball. Homosexual men performed significantly worse than heterosexuals on this task and were not significantly better than women. Kimura does not mention whether they tested the fine manual skills of the homosexuals.[22]

Another difference that is unrelated to cognitive abilities but is consistent with reduced early exposure to testosterone is the fact that children with gender identity disorder, many of whom grow up to be homosexual, are judged to be better looking than average. Their faces are seen as more youthful, and it seems unlikely that this could be attributed to taking more care of their physical appearance because it is a matter of bone structure.[23] Women are also judged to be more attractive than men, and their beauty relies partly on creating an impression of youthfulness. Thus, extremely attractive women have some of the facial proportions of young girls. Since beauty emerges in young children, it seems that traits promoting physical attractiveness may be expressed quite early in development, possibly before birth. Like all sexually dimorphic traits, the difference in physical attractiveness between sexes is presumably due to the impact of sex hormones. The physical attractiveness of homosexual men could be an echo of reduced exposure to testosterone early in development. Testosterone has the effect of elongating bones, making the face appear more rugged and mature, so that decreased early exposure could make them seem more boyish in appearance.

Researchers in brain biology have produced a great deal of compelling evidence that sex differences in aptitudes, interests, and behavior, including romantic behavior, emerge early in life due to the influence of sex hormones on the developing brain. These differences have been quite controversial because they are rather inaccessible to study. Even so, men and women clearly differ behaviorally in ways that would have helped our ancestors survive and reproduce. Thus, it is likely that the sexual attraction of most people to members of the other sex is due to the influence of sex hormones on the brain before birth, although these brain mechanisms are not triggered or activated until there is a spurt in sex hormone production at puberty.

Romance is thus etched in the biology of the brain just as surely as circuits are etched on computer chips by Intel. The main difference is that humans and other animals are in a constant state of change, whereas electronic circuits remain the same. When human lovers meet, or when pigeons court, a similar drama of hormones and neurotransmitters is evidently being played out in their brains, getting them in the mood for mating and living together.

THE CHEMISTRY OF LOVE

Sleeplessness, poor appetite, excitement, cravings, and euphoria are classic symptoms of the romantic conception of being in love, and people in most societies interpret them in this way. They are also classic effects of stimulant drugs such as cocaine and amphetamines. The similarity can be explained in terms of the chemical effects of falling in love.

Is lovesickness a physiological state analogous to drug intoxication? Researchers have discovered a stimulant neurotransmitter, phenylethylanine (PEA), that is elevated in people who are in love. This is apparently released from the moment of contact with the object of one's affection, which might help to explain the frequently reported phenomenon of love at first sight.[24] Cupid's arrow flies and a whistle blows in the PEA factory. The results are googoo eyes and silly smiles flashed at the attractive member of the opposite sex.

PEA stays high for the first two or three years in most relationships. As the Cole Porter song says, the "kick" is from the loved one, not from cocaine. The high does not last, however. Many people who fall in love and marry can't sustain that high. According to anthropologist Helen Fisher, there is a peak in the divorce rate in the fourth year of marriage throughout different countries around the world.

Fisher claims that the chemistry of relationships arises from an ancestral past in which marriages lasted only long enough to raise a child through the most difficult period of dependency, i.e., the first few years during which the child is breast-fed and must be carried around by parents, but this is controversial.[25]

In our own society and in many others, however, marriages can be lifelong. Some people are as monogamous as swans. The early excitement is replaced by calmness, intimacy, and a sense of well-being. Psychologists refer to this as *companionate* love, distinct from the passionate love of newlyweds.

At the risk of misunderstanding, romantic relationships can be compared to drug habits. In the early days, they are like amphetamines or cocaine, producing an intense feeling of euphoria that is celebrated in song and verse. The stimulant rush makes us excited and happy, which motivates us to be around the other person as much as possible. Yet this euphoria, as already pointed out, is not to last.

One of the great ironies of drug addictions is that some drugs that produce an intense pleasurable rush, like amphetamines, are not very addictive. The most addictive class of drugs is the opiates that produce feelings of euphoria and calmness, rather than euphoria and excitement. Recent research has demonstrated that all kinds of affectionate relationships increase the supply of oxytocin in the blood for men and women. Oxytocin is an opiumlike hormone (strictly speaking, a polypeptide) produced by the posterior pituitary gland at the base of the brain. It has come to be called the cuddling hormone because it is produced when people cuddle. Oxytocin is produced in mothers when they are breast-feeding and plays an important role in the milk letdown response. It is also released during sexual behavior and evidently plays a role in sexual pleasure.[26]

Oxytocin has the ideal chemical properties for getting people hooked on each other, just as heroin addicts get hooked on opiate drugs. Long-term relationships may not be at a high pitch of excitement for most of the time, but it would be a mistake to underestimate their chemical strength. Husbands and wives may not realize how much they love and rely on each other until they are separated for a prolonged period of time. Then they become moody, restless, and irritable, like opium addicts denied a fix. This fix has nothing to do with sexual behavior as such, but comes from the quiet pleasure of being in each other's company, just as might be true of a pair of pigeons cooing together beside their nest in the evening. Oxytocin casts its soothing spell to keep romantic partners together over the long haul. It is probably helped by a class of brain neurotransmitters, known as the endorphins, or natural opiates, that relieve pain and promote calmness.

What has been said above implies that men and women are similar in their vulnerability to the chemical addictions of love, and this is largely true. When you look beneath the surface, nature is surprisingly androgynous but there are some differences. Men—and male organisms, more generally—are quicker to become infatuated than women are. They are more drawn to the sheer excitement of sexual behavior.

The same is true of monogamous prairie voles, according to an amusing experiment by University of Maryland zoologist Sue Carter and her colleagues. Carter had the voles engage in the stressful experience of swimming for three minutes before encountering a member of the opposite sex. Once the female voles had been liberated from their bath, they lost interest in mating and scurried away. For the males, on the other hand, the females were unusually attractive and they bonded more rapidly with them than would have happened in the normal course of events.[27]

Surprising though it may seem, many conceptually similar experiments have been performed on human beings, almost all unfortunately using just male participants. Social psychologists Donald Dutton and Arthur Aron initiated this sequence of human experiments in the early 1970s by arranging an ingenious test of the hypothesis that excitement gets people's romantic juices flowing, so to speak. Their study was conducted on two pedestrian bridges at a tourist destination in British Columbia. One bridge was stable and close to the ground. The other, the Capilano Suspension Bridge, was hung 250 feet above a rocky gorge. Extending for 450 feet and swaying perilously in the breeze, it was quite an ordeal to cross, particularly if you happened to be afraid of heights. Either a male or a female researcher met unaccompanied males whenever they crossed either of the two bridges. The researchers asked them to participate in an experiment and obtained answers to a few questions. Participants were then given the researcher's phone number so that they could ask for more information about the study. Men who had been met by female researchers on the swaying bridge were more likely to call than men who had met the same women on the stable bridge. Evidently for them, as for the male voles, the stress of crossing the frightening bridge had made them more excited by the female researcher they had encountered. Despite many subsequent studies designed to challenge this interpretation, the original study and its interpretations have held up.[28]

It is interesting that homosexuals fall in love in much the same way as heterosexuals. There is no denying that gay men and lesbian women can be just as romantic and just as obsessed with each other as heterosexual couples. Even though homosexuals are excited by members of their own sex, the brain chemistry that sets their heart racing and makes their knees go weak is evidently the same.

While some people may be offended by the notion that our romantic lives can be reduced to chemistry, it is really no more offensive than the uncontroversial notion that our bodies are composed of atoms. Chemical reduction is a scientific method of explaining psychology and behavior. Knowing that our romantic feelings are predicated on chemistry subtracts nothing from our experience of them—any more than knowing that sexual intercourse is a question of physics, and that gravitation detracts from connubial bliss.

Falling in love may be a very complex phenomenon for human beings compared to other species. When a female gypsy moth is ready to mate, she releases a sex-attractant pheromone. Males are fantastically sensitive to this substance, responding to a single molecule and being able to detect a female from a distance of more than a mile if the wind is blowing in the right direction. Recently, scientists have suggested that human pheromones could provide a direct route by which men and women may affect the chemistry of each other's brains.

PHEROMONES

A pheromone is a chemical realeased into the air by one member of a species that has the function of changing the behavior, or internal physiology, of another. Pheromones may be detectable as odors, or they may pass through a primitive sensory organ (the vomeronasal organ) that is present in the human nose and is capable of detecting chemicals of which we are unaware. However, in humans, this is a matter of ongoing research and remains controversial.

The first indication that there are human pheromones was the finding that girls sleeping together in a dormitory have synchronized menstrual cycles. Subsequent research showed that the underarm secretions of one woman could be used to drive the menstrual cycles of others. A related finding is that when women begin sleeping regularly with a man their menstrual cycles become more regular and they ovulate more often. It is a mistake to read too much into such findings, especially when we do not understand what practical consequences, if any, they might have had for our primeval human ancestors. Yet they at least show that the chemical channel, which is so critical for the reproductive behavior of most mammals and other species, is still functional in humans.[29]

In a fascinating recent study, Winnifred Cutler and her colleagues at the Athena Institute for Women's Wellness in Chester Springs, Pennsylvania, tested whether a synthesized version of a human male pheromone would affect men's ability to sexually arouse women, as reflected in increased rates of sexual intercourse.[30] Participants in the study wore an aftershave that either contained the possible pheromone or a placebo that could not affect sexual responsiveness of the men's romantic partners. Participants did not know which they were getting. For most of the men wearing the pheromone there was no change in frequency of sexual intercourse, but eight of seventeen in the group reported an increased frequency of intercourse. This was not simply due to the expectation that being in the study would enhance their sex lives, because only two of the twenty-one men in the placebo group had an increased frequency of intercourse over their regular (baseline) level. The difference between the two groups was statistically significant, allowing the researchers to conclude that secretions produced in men's armpits *can* affect sexual desire in women. They do not appear to have affected sexual desire of the male subjects because, for example, their frequency of masturbation did not change. It is possible, however, that the effects were produced by increased self-confidence of the male subjects. The results are preliminary and they involved a small number of people, so that replication is necessary before they can be taken seriously.

Of course, there is no guarantee that such knowledge will always be used for good intentions. Unscrupulous entrepreneurs are currently marketing, over the Internet, possible human pheromones that can be worn by a man "to make her lose control of her animal instincts." Some scientific evidence suggests that pheromones can be important to human sexuality, that they may unobtrusively influence perceptions of sexual attractiveness. Yet there is little consensus about what these substances are, and there is little research on their actual behavioral effects. A more responsible approach is taken by Erox Corporation, the manufacturer of *Realm* fragrances for men and women, which contain possible pheromones, but are marketed as boosting self-confidence rather than promoting sexual intercourse.

Internet marketing seems unduly optimistic because pheromones can only work at close distances due to the general insensitivity of human chemical senses compared to those of other mammals. One suspects that eager purchasers who are relying on this product alone will experience disappointment in their attempts to bypass the usual channels of evaluation by which potential sexual partners size each other up. Their problem is that visual and social screens will drop before them like so many portcullises. For a man's pheromones to work, he must first get close to the woman, and if he is allowed to enter her personal space, he may not need them.

Natural selection has designed the brain to produce an appropriate chemical response to a potential mate and it is even possible that this response is affected by chemical signals in the air. Selection has also designed the visible signals that trigger such a response by sculpting the human face and body. Differences between the bodies of men and women are accentuated at puberty. These differences are the basis of sexual attraction and allow people to select their mates.

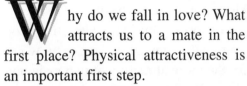

Chapter Two

Physical Attractiveness and Sex Signals

Why do we fall in love? What attracts us to a mate in the first place? Physical attractiveness is an important first step.

Favoring attractive people is a deeply ingrained human tendency. Moralists often decry our society as shallow and excessively concerned with physical appearance, but beauty is celebrated in every society. Among many subsistence peoples, physical appearance is a constant concern. New Zealand's traditional Maoris tattoo their entire faces. Some African women expand their lips with wooden discs the size of tea plates. Others stretch their ear lobes by several inches using heavy weights. Such practices enhance physical attractiveness in other societies, although we Westerners may find them distinctly unappealing. Despite differing opinions about the impact of such artificial manipulations of bodily appearance, all societies agree about which people are naturally beautiful (i.e., before the artificial modification) because evolution has built some esthetic preferences into all of us.[1]

Attractive people are clearly in great demand as dating partners and as spouses, but they enjoy many other advantages because of an irrational preference for physical attractiveness built into us all. Employers like to hire attractive people; juries are less likely to find them guilty of a crime; teachers more often excuse attractive children for their bad behavior. To top it off, attractive people are blessed with better health.[2]

Favoring physically attractive people is unfair and often wildly irrational. After all, what does a woman have to gain from voting for the more attractive of two male political candidates? What does a man have to gain from helping an attractive stranger whose car has broken down? Once he has solved the problem, she roars off into the sunset, never to be seen again. Favoritism is automatic and irrational. We favor attractive strangers because natural selection builds the tendency into our brains.[3] Such a preference would have helped our ancestors to choose their mates well. There is nothing specifically human about the power of beauty.

SEXUAL SELECTION

Consider the evolution of esthetic preferences among animals, first discussed by Charles Darwin (see fig. 1). These can be understood as due to a special type of evolution, known as *sexual selection*. Darwin discovered *natural selection*, the process by which survivors pass on their successful traits to offspring. Birds of the Galapagos Islands, where he studied, are known as Darwin's finches. Ground finches (*Geospiza magnirostris*) die in large numbers during drought years because of food scarcity. Finches with large beaks have an advantage because they can break open the tough casings of hard

Fig. 1. Charles Darwin depicted in old age is himself "ornamented" with the sexually selected male trait of beardedness (see text for explanation). (Reproduced from the Collections of the Library of Congress, LC-USZ62-123826)

seeds that others cannot. Since only the large-billed birds survive, their young inherit big beaks. Evolutionary change can thus occur in a single generation, much more rapidly than Darwin suspected. During years of abundant food, however, having a smaller beak is an advantage because small food items can be more easily manipulated—evolution then shifts the population back in this direction.

Darwin recognized that there are two kinds of evolution. One solves the problems of survival posed by the environment, such as the finches having the proper type of beak to exploit the available seeds. This is called natural selection. The other type of evolution solves the problem of acquiring mates, and is known as sexual selection. Darwin explained the gigantic colorful tail plumage of male pheasants and peacocks by arguing that the "ornamentation" of the males existed to seduce females.

A show requires an audience. Peahens (female peacocks) are discerning critics. They almost never mate with the first male they encounter, preferring to inspect several before making their choice. Peahens prefer to mate with the most gorgeous peacocks because they are the ones with the best genes—not that peahens know anything about genetics, of course. Progeny of a union between a peahen who is turned on by male beauty and the object of her attraction have two distinct advantages. The male offspring will be capable of seducing females, which means that they will sire many chicks. The female offspring will be attracted by sexy peacocks, which means that they will choose mates whose genes help the offspring to survive.

In the natural world, there is thus no division between beauty and function. The sweet odor of flowers is an invitation to the insects to pollinate them. Their bright colors allow bees to systematically visit a particular species at a time when it is putting out nectar through simple association learning. An Australian orchid mimics the appearance and odor of a female small black wasp, inducing males to copulate with them. This lodges their pollen upon the wasp's bristly body so it can be transported to other plants to fertilize them.[4]

Since the peacock spends much of its time fanning its tail

feathers and displaying them in the direction of the female, thus getting her attention, the colossal train was obviously fashioned by natural selection to appeal to peahens. Why peahens should choose to mate with peacocks having such gigantic feathers to drag around after them was the subject of much controversy among evolutionary biologists. The seemingly needless burden of a long tail is a drain on energy resources when food is scarce and it makes the peacock more conspicuous to predators; theoretically, long-tailed peacocks could be more easily caught and killed.

SEXY SONS VERSUS GOOD GENES

There are two schools of thought about the benefits to peahens of opting to mate with gaudy males. The first, called runaway selection, says that bright plumage is like a fashion fad that takes off for no very good reason and goes to an extreme. Suppose that ancestral peacocks began with the same drab plumage as the females. Then a mutation arose that made males more colorful. Suppose further that the colorful males were attractive to at least some females. The offspring of these unions would have consisted of both gaudier males and females preferring such gaudiness. The greater the number of females with this preference, the greater the advantage to males of being colorful. Over generations, the male trait could thus become more exaggerated. Moreover, as the reproductive success of the colorful males increased, females would have an additional reason for mating with them. By mating with such highly attractive males, the females would produce sexy sons, thereby maximizing their number of grandchildren and winning in the great competition to reproduce that is the driving force of evolution. This is so because individuals that fail to reproduce have their genes taken out of circulation, and their unsuccessful traits then disappear. This is the mechanism of natural selection. Dullness in the plumage of peacocks was a trait that could not compete with gaudi-

ness and was removed by natural selection. Peahens preferred
males with the most colorful plumage. Over evolutionary time, the
length and colorfulness of the peacock's train got exaggerated
through runaway selection, a process halted only when its survival
costs canceled out the reproductive benefits of a large colorful tail.

The alternative explanation to runaway selection is the *good
genes theory*. The good genes theory argues that while colorful
plumage may seem arbitrary, it is a useful outward and visible sign
of inward genetic advantages. Brightly plumaged males could be
"advertising" an inherent resistance to parasites, for example, or a
superior immune system that helps them to fend off viral infections.[5]

English biologists Bill Hamilton and Marlene Zuk tested the
good genes theory by comparing different bird species. They discovered that species with the brightest coloration of the feathers
and most pronounced musical songs also had the most problems
with blood parasites.[6] Because parasites were such a problem,
males who could demonstrate resistance to them by being especially colorful and singing vigorous songs would be more attractive
as mates. This argument makes the assumption that, within a colorful species, individuals that carry a heavy parasite load will have
difficulty in producing bright plumage colors. Females selecting
males on the basis of their colorful feathers would thus be acquiring
genes for resistance to the parasites that beset their species.

Even more convincing support for the good genes theory comes
from evidence, in an experiment conducted in England, that the offspring of peacocks with larger eyespots in their tail feathers grew
faster and lived longer in the open. We do not know why they lived
longer, but the essential point is that females choosing the gaudier
males enabled the selection for biological superiority, as reflected
in better survival rates. Other interpretations, such as the possibility
that brighter males had a better chance of avoiding being run over
by motorists, are possible but unlikely.

Mating with attractive males would give females offspring that
were better at surviving as well as sons that were more attractive to

the next generation of peahens. *Peahens do not have any grasp of the connection between physical attractiveness and good genes, of course.* Their reactions to male beauty are automatic. They do not have to *understand* beauty. All they have to do is get turned on by it.[7]

Charles Darwin noted that male animals are generally gaudier and more interesting to look at than females. Since females are more in demand as mates than males are (because they contribute more to offspring by laying an egg or carrying a fetus to term), they can be a lot choosier. They can afford to sit back as if to say, "I will join my genetic future only with the strongest, healthiest male." Males are thus confronted with the task of winning over the female. This is why it is generally the male that appears gaudy and attractive. Its fancy wrapping is there to "sell the product." In the language of economics, the male bears the burden of advertisement.[8]

Among humans, both sexes agree that women are the more physically attractive, or the naturally more ornamented sex. For example, women participating as subjects in psychological research receive higher ratings on physical attractiveness scales from both sexes. Women also spend a great deal more time and effort on their appearance and dress, suggesting that these are of greater feminine concern. Such behavioral enhancement of physical appearance bears analogy with the peacock strutting his stuff. It is fascinating that women are generally much more interested in their own appearance than men are—regardless of their economic power compared with men, their level of physical attractiveness, or how difficult it is to find the right marriage partner. This feminine concern provides an important insight into the ground rules of human reproduction. If women are bearing the burden of advertisement, then men are in a stronger bargaining position relative to most other male mammals and birds.[9]

THE GROUND RULES OF HUMAN COURTSHIP

Understanding courtship in other species illuminates differences between the sexes and explains why there are conflicts of interest between men and women in courtship and marriage. The conclusion that women are sexier in appearance than men suggests that women are competing among themselves for access to men. Yet this conclusion does not ring true in the real world. According to anthropologist Don Symons, of the University of California at Santa Barbara, sexual intercourse is everywhere a female favor granted to men. Any woman, whatever her looks, can succeed in becoming pregnant.[10]

Why then do people of both sexes consistently rate women as more physically attractive? They do not need to be physically attractive to have sex with men. They are obviously not competing over opportunities for sexual intercourse. Instead, they are competing to *marry* desirable husbands who can help them to raise children.

Men's bargaining position is based on social status and wealth. In subsistence hunter-gatherer societies, such as the Siriono of Brazil, a man's sexual attractiveness to women is based largely on his reputation as a successful hunter.[11] In modern societies, women are more interested in a man's education and income level than his hunting ability, but this concern represents the same underlying need to find a mate who will be a good provider of food and other economic goods. Even though modern women sometimes earn more than most men, their evolved psychology has not changed.[12] They are still attracted to successful men. Some striking examples of this phenomenon in modern sexually liberated women are presented in chapter 4.

In early subsistence societies, as well as more recent ones, women were constantly pregnant, breast-feeding, or caring for children—or all three. This would have interfered with their ability to work and acquire surplus food or property. Today, the birth rate is much lower due to the use of effective birth control techniques. Moreover, children spend the day in daycare or with babysitters, which frees their mothers for full-time occupations. Successful rearing of children among our hunter-gatherer ancestors was a cooperative enterprise in which men contributed to feeding, sheltering, carrying, protecting, and caring for their offspring. The critical importance of fathers for the survival of their children is demonstrated by the Ache of Paraguay, who are more than twice as likely to die during childhood if they lose their father[13] (see chapter 4). Women, in general, have evolved to compete for husbands with social status and wealth because these are reliable cues to the ability to protect and care for children.

Physically attractive women (as assessed from high school yearbook photographs) are much more likely to marry. They also marry up the social ladder, finding husbands that are wealthier than their parents;[14] the same, however, is not true of men. Physically attractive women move up into wealthy elites through marriage, while physically attractive men do not.

Peahens are very plain in appearance, suggesting that peacocks are not drawn to the physical attractiveness of mates. The evolutionary logic behind this male indifference is that peacocks invest so little in their offspring they do not have to be selective in the choice of a mate. Every female that they mate with will likely increase their reproductive success.

The corresponding lower physical attractiveness of men compared to women suggests that women are not as concerned about the physical attractiveness of their mates.[15] Does physical appearance really not matter to women in their selection of a spouse? Why not ask women what they consider important in a potential husband? This straightforward approach is fraught with unexpected problems. What if people have limited insight into their own actions? What if they are telling white lies to make themselves look good or to provide what they believe is the "correct" answer to the question?

When asked, women and men agree that personality is all-important. They want a spouse who is kind and understanding, who is intelligent, and has a sense of humor. Their actions belie their words. In a classic study conducted at the University of Minnesota, Elaine Walster and her colleagues organized a freshman dance.

They told couples that a computer program would match them on the basis of personality. In reality, men and women were randomly paired up. The researchers were interested in which personal characteristics would make people want to have a second date with their dance partner. To their astonishment, the researchers found that no personality measure predicted partner desirability. The only predictor was physical attractiveness. Both men and women wanted to be with the beautiful people! This shows an astonishing split between people's stated motives and their actual behavior. Either we are blind to the birdbrained mechanisms that control interpersonal attraction or we are unwilling to admit to them.[16]

More recent research shows, however, that occupational status can overwhelm the positive effects of physical attractiveness. If an attractive man puts on a business suit and wears an expensive watch, women find him far more desirable than the same man dressed in a Burger King uniform.[17] Men are far less picky about the occupational status of women they are willing to date.

The discrepancy between what people say and what they do may have an innocent explanation. It could be that admirers view attractive people as having desirable personality traits. This phenomenon is known as a *halo effect*. There are strong halo effects for physical attractiveness. In other words, physical attractiveness creates a favorable personality impression, and we imagine we like the person for his or her personality. The positive first impression that attractive people make on others means that they have an advantage in getting to know people, which helps them to project a positive image of their personality. Whatever else is said about the University of Minnesota study, it finally dispels the myth that women do not care about physical attractiveness in men. Appearance may not be as important as social status *but once the status hurdle has been crossed*, physical attractiveness may be the single most important influence on women's dating choices.

To an evolutionist, saying that women are indifferent to the physical attractiveness of their mate is equivalent to saying that they behave as if the genetic quality of a man is unimportant. Such care-less choices could not have survived the ruthless forces of natural selection. Women who paid more attention to the genetic quality of their mates would have left more surviving descendants to inherit such fastidiousness. Like peahens, they are turned on by the orna-mentation of their mates, even if this is not obvious on the surface.

WHAT WOMEN LIKE ABOUT MEN'S BODIES

Among our primate cousins, males have many facial and bodily displays, including beards, that are intimidating to other males but attractive to females. Beard growth comes with the spurt in testos-terone production at puberty. In that sense, facial hair is an obvious sign of reproductive maturity. Testosterone level also determines sex drive and sexual potency in men. This means that male facial hair is an outward and visible sign of reproductive capability. Although men in some societies produce very little facial hair com-pared to Europeans, they almost always have more than women do.[18] This may seem obvious, but it is a necessary part of the argu-ment that male facial hair is a sexually selected trait.

Are beards attractive to women? One researcher simply asked women whether they found beards attractive. The overwhelming majority emphatically denied that they did, but we know that people's verbal or written responses on such matters are not always reliable.[19]

During the 1970s, social psychologist Robert Pellegrini of San Jose State University studied reactions to bearded undergraduates without mentioning beards. Some of these bearded students were strapped for cash and willing to shave for $20. They shaved their beards in a stepwise fashion, fully bearded, moustache and goatee, mustache only, and clean-shaven. Researchers photographed them at each stage. They then showed the pictures to others. The viewers evaluated the person depicted in the picture on many personality traits. Both male and female evaluators considered photographs depicting the full beards as most attractive. They also rated them more favorably as to maturity, confidence, dominance, courage, industriousness, masculinity, and creativity. Each of these qualities would go on a shopping list of desirable qualities in a husband. In general, the more hair left on the face, the more people liked it.[20]

Pellegrini's results have been replicated many times since, showing that they are not peculiar to the late hippie era of the 1970s, when beards and mustaches were still quite popular. One study, published in 1986, found that ratings of physical attractiveness increased steadily with the amount of facial hair in Identi-kit pictures.[21] Another, published in 1989, reported that bearded men were perceived by both men and women as more masculine, stronger, and more dominant.[22] A third, published in 1990, looked at the effects of facial hair on perceptions of job applicants by 188 personnel managers.[23] Managers saw bearded applicants as more socially and physically attractive, more self-assured, more competent, and more personable. A Swedish study, published in 1994, found that wearing a beard conveyed impressions of originality and goodness.[24] Swedes also see bearded faces as better looking, more masculine, and more congenial.

Beards are attractive because they create an impression of health and vigor, or biological fitness. People describe bearded men as more active and potent, reflecting the biological association between testosterone levels and virility. Bearded faces are also rather intimidating, in the sense that they convey an impression of dominance or high social class, as in the depictions of Moses and other biblical leaders. Women are interested in husbands who have high social status. Having a beard thus helps men to seem more desirable on most of the qualities that women use to select a husband.

Such findings are remarkable given that they were produced in the 1970s, when most men did not actually wear beards or mustaches and women's favorite actors, for example were—and still are—are mostly clean-shaven. In fact, many people today associate wearing beards with homelessness. It seems that our visceral reactions to beardedness can overcome such unfavorable stereotypes. The social significance of facial hair must change greatly over time to explain fluctuations in beard fashions. One interesting hypothesis is that when marriages are unstable, as is currently the case, women are attracted by clean-shaven men because their romantic intentions are easier to discern when their facial expressions are not hidden behind a mask of hair.[25]

Another characteristic that sets some men apart from others is height. Clearly, height conveys an impression of social dominance. In many preindustrial societies, people call village leaders "big men," and they are often tall. Height goes along with elevated social status in modern societies as well. Taller men have advantages in getting hired and are more likely to be promoted in organizations, resulting in a salary advantage. The relationship between height and social status is quite marked. For example, in England, a very class-conscious country, people with lower occupational status are several inches shorter than those with higher status. This difference could be due to nutritional inadequacy among the poor, or it might be partly genetic in origin. An interesting exception is provided by the Pygmy hunters of central Africa. Among the Pygmies, tall men are considered awkward and are believed to be poor hunters. It is possible that height is a genuine disadvantage for moving through the dense undergrowth of rain forests.[26]

Given the nearly universal association between height and social status, it is no surprise to discover that women prefer to date tall men. They want to go out with a man who is at least four inches taller than themselves. This principle is called the cardinal rule of dating, and it is very rarely violated. In one study of 720 married couples, for example, there was only one couple in which the woman was taller than the man—although there should have been twenty-nine such pairings based on the distributions of height for men and women (men being five to six inches taller) if pairing were random.[27]

While men have little consistent preference for height of women, women tend to dismiss short men as possible dating partners. Exceptions may be made for exceptionally prominent men like Napoleon Bonaparte, who measured five feet and two inches, but had no shortage of lovers despite his diminutive stature. Being more popular as dating partners, tall men go out more often and with more different women, presumably because there is a larger pool of women for whom they measure up to the cardinal rule of dating. Height is only one dimension of body build, and it would be strange indeed if women were indifferent to other bodily indicators of the ability to survive and reproduce.

Although women are attracted to muscular men, they do not like heavily muscled bodies of the Mr. Universe type. This is probably a sensible preference because the huge muscles of body builders are unnatural. They are often facilitated by use of anabolic steroids, which can have devastating health consequences. Even though our remote ancestors obviously did not use steroids, too much muscular tissue would have strained their cardiovascular system, making them vulnerable to heart disease, for example. Women have apparently evolved to prefer the body build of an effective hunter: strong enough to subdue prey, but not too heavy to impede mobility on long hunting trips or place an undue strain on the cardiovascular system.[28]

American women favor men with broad shoulders and tapering torsos. In one study where participants rated the attractiveness of various line-drawing silhouettes, the breadth of the shoulders was much more important than any other dimension.[29] However, if men are depicted with bloated stomachs, the bulging stomach becomes most important—as a repellant trait!

Symmetry, or the exact match of the left and right sides of the body, is important to the attractiveness of both sexes. Both sides of the face should be exact mirror images of each other. Kevin Costner has a far more symmetrical face than Lyle Lovett and thus Costner is considered better looking. Careful investigations by biologist Randy Thornhill and his colleagues at the University of New Mexico at Albuquerque have shown that people with symmetrical faces generally have symmetrical bodies.[30]

Bodily symmetry is an esthetic cue used to assess the biological fitness of potential mates among other species. For example, research on swallows, which have forked tails, has shown that females prefer to mate with males having symmetrical tails. Asymmetry is caused by interference with normal development, which might be due to poor nutrition early in development or might reflect the impact of viruses. Symmetrical animals have superior biological qualities either because they experienced a favorable early environment or because their immune systems were effective at warding off viruses and other pathogens. Swallows and people attracted to mates with symmetrical bodies acquire a superior immune system for their offspring.[31]

This explains why women should be attracted to highly symmetrical men. Thornhill and his colleagues have discovered that symmetrical men have more sex partners, and even that women get more excited during intercourse with these physically attractive men. Symmetrical men produce a pheromone (or airborne hormone) that is more attractive to women than the secretions of less symmetrical men, suggesting that women's attraction to men is based on assessment of biological fitness through different sensory channels.[32] Men's biological quality declines with age, which is reflected in declining facial symmetry, for example. This may have important implications for offspring. Thus, declining sperm quality of older men increases the risk of Down syndrome and other chromosomal disorders.

THE DOUBLE STANDARD OF AGING

Men's physical attractiveness to women declines with age, but the decline is generally less steep than that of women to men. In what might be called the second cardinal rule of dating, men want partners who are a year or two younger than they are, while women, in general, want to date older men. As men age, they want women who are increasingly younger than they are. A man of forty, for example, is likely to want a partner who is ten years younger. Why?[33]

The most fundamental reason relates to limitation of women's ability to conceive children with advancing years. Fertility reaches a high point in the early twenties and stays on a plateau until the age of thirty-five, after which it declines sharply. Natural selection would have caused men to select fertile women as wives since those who were attracted to women over fifty would have left no offspring to carry on their unusual taste. However, men see women as more attractive at twenty than at forty. This is right at the beginning of their most fertile phase in the life span.[34]

Men are thus most attracted to women who are at the beginning of their reproductive career. If a man marries a woman of this age, then he has the potential of giving her *all* of her children and thereby hitting the reproductive jackpot. Natural selection has thus favored men who are attracted to younger fertile women rather than older fertile women. For this reason, the perception of youthfulness is critical to the physical attractiveness of women. This helps explain the success of the cosmetics industry, as women attempt to conceal signs of aging and try to appear younger and more attractive.

Men reach the peak of their physical attractiveness to women in the late teens or early twenties. However, as they grow older, they acquire social status and wealth, which enhances the value of the overall package as far as a marriage partner is concerned. Although men deteriorate with age, their physical appearance is less critical to their overall attractiveness. One important cue to feminine youthfulness that plays an important role in women's physical attractiveness is their bodily shape.[35]

SEXUAL SELECTION AND THE HOURGLASS FIGURE

The body shapes of men and women are sexually selected traits, analogous to the plumage of the peacock. Strange as it might seem, this conclusion is supported by much compelling evidence. To begin with, feminine curves emerge around puberty, just like the colorful train of the peacock (see fig. 2). They are produced by the same mechanism, a surge in production of sex hormones. A surge of the sex hormone estrogen stimulates the filling of fat cells located away from the waist. The "loudness" of the signal (i.e., the size of the sex difference) diminishes with age. In the same way, the greater height of men, their broader shoulders, their deeper voices, and their greater upper-body musculature are due to the growth spurt produced by a surge in testosterone production at puberty.

Both sexes agree that women with "hourglass figures" are sexy and attractive (see fig. 3).[36] This contrasts with the attractive male body. In sexy men, there is little difference between the hip and waist dimensions, the torso is moderately muscled, and the shoulders are broad.[37] The attractiveness of an hourglass figure for women is a constant across cultures and across time, although the amount of curvature considered desirable varies greatly in different countries and at different times within a society.

Scientists assess the curvaceousness of the human body using a statistic known as the waist/hip ratio. A small waist/hip ratio is equivalent to a highly curvaceous body. Highly attractive women, such as Miss Americas, have a waist/hip ratio of about 0.67 (the ratio produced by a waist of 24" and hips of 36", for example). The normal range for women is 0.67–0.80, whereas the normal range for men is 0.85–0.95. Lack of an overlap between the male and the female range means that body shape is a highly predictable sex difference.

The intensity of the signal (i.e., the size of the sex difference) declines with age due to a change in hormone production. If you are on the beach and spot a couple strolling away from you in the distance, distinguishing the silhouette of the man from the woman will be very easy if the couple is in their twenties, but much more difficult if they are in their fifties.

One distinguishing characteristic of the peacock's tail is that it interferes with movement. Similarly, storage of fat away from the waist is not mechanically efficient. It makes more sense to store fat close to the center of gravity, in the abdomen. Highly curvaceous

GODEY'S FASHIONS FOR FEBRUARY 1865.

Fig. 2. Women's bodies were sculpted by the same process as the peacock's colorful tail feathers. These fashions of 1865 also seem inspired by the peacock. (Reproduced from the Collections of the Library of Congress)

women are at a distinct disadvantage in sports and rarely win Olympic medals—for example, in events requiring agility and speed, such as basketball and running. This is not to claim that curvaceous women cannot be very athletic. Some are, but when they compete at the highest level, they experience a mechanical disadvantage because weight stored away from the center of gravity introduces turning forces that use up energy.[38]

Just as peahens are attracted to an extremely colorful mate, so extremely attractive women are at an extreme of the range for curvaceousness. Beauty contest winners cluster at the curvaceous extreme of 0.67 compared to the normal range for women of 0.67–0.80.

REPRESENTATION OF A BALL DRESS.

Fig. 3. Waists were tightly cinched to exaggerate the female form and enhance sexual attractiveness, as in this print from 1854. This extreme is not considered attractive today. (Reproduced from the Collection of the Library of Congress, LC-USZ62-098063)

Perhaps the most important and compelling point of similarity is that curvaceous women, like showy peacocks, have superior immune systems. According to recent research of Devendra Singh, an evolutionary psychologist at the University of Texas at Austin, college-aged women and men agree that curvaceous women (whether of normal weight, underweight, or overweight) are more attractive, healthier, and more capable of producing children than less curvaceous women. What makes these findings really interesting is that they are borne out by medical data. Women with low waist/hip ratio (i.e., with a curvaceous build) not only have less difficulty becoming pregnant, they are also healthier in terms of a lower incidence of many illnesses. Women with relatively noncurvaceous bodies are at a higher risk for gall bladder disease, some cancers, diseases of the heart and circulatory system, and for diabetes. (It is important to realize that a curvaceous body is different from an obese one: curvaceous women store fat away from the abdomen whereas obese women usually have thick fat deposits around their middle which pose major health risks.) Noncurvaceous women are also more prone to behavioral disorders such as anxiety and drug abuse. (It is true that some drug addictions can cause people to lose weight, which might make them less curvaceous, but the finding applies equally for alcoholism, which can have the opposite effect.) Less curvaceous women are more likely to be admitted to psychiatric hospitals for depression and other psychopathologies. They also have higher mortality rates. The health consequences of body shape in men have received less attention because the waist/hip ratio is not routinely measured for medical records and is thus unavailable to researchers. Women are at their most curvaceous in early maturity, and one reason that men are attracted to women with sexy bodies is that this is a cue to youthfulness. Exceptionally attractive women have youthful facial dimensions that make them seem more attractive than they really are.[39]

EXAGGERATING YOUTH

Men are very sensitive to age cues, since a woman's age places limitations on her ability to produce children. It is true that men may
be motivated by opportunities for sexual intercourse rather than
opportunities for reproducing, but natural selection has designed
them to want sex with fertile women. The sex difference in the
importance of youthfulness to physical attractiveness explains why
women are much more interested in using makeup to make them
seem young and healthy than men are.[40]

Highly attractive women, such as film actresses, often preserve
their good looks into old age. The impression of youthfulness is so
powerfully conveyed by the design of their faces that seeing them
as old is difficult. When you see Candice Bergen, now well into her
fifties, for example, your response is likely to be, "What a stunningly beautiful woman!" rather than "Candice seems to be well
into her fifties."

Highly attractive women's faces mimic some facial proportions
normally found only in young girls. This is not just true of our own
culture with its exaltation of all things youthful, but in Japan and
other countries.[41] The extent of this phenomenon can be grasped
from the fact that attractive male faces tend to have large chins, a
feature associated with age and maturity, whereas attractive female
faces have reasonably small chins and therefore resemble the faces
of children. Similarly, small noses contribute to the attractiveness
of women's faces, but the size of the nose is unrelated to attractiveness of men.[42]

Other youthful traits that people see as attractive in women (but
not men) in different cultures include dainty hands and small feet.
These are not just smaller in women because they have shorter stature,
but they are proportionately smaller. Small feet were such a critical
feature of attractiveness in China that parents kept their daughters'
artificially small through the hideous practice of foot binding. From a
biological perspective, small chins, feet, and hands are attributable to

low levels of testosterone, which promotes bone growth. Since testosterone reduces female fertility, they are thus an outward sign of a hormonal profile conducive to high fertility in women.[43]

Victor Johnson, an evolutionary psychologist working at the University of New Mexico at Las Cruces, has collected other evidence that shows the extensive effects of sexual selection on the female face. Johnson and his associate used a computer program that allowed people to "evolve" their perfect faces over many "generations." One striking aspect of the perfect faces was that many dimensions were typical of much younger faces. Although people estimated the age of the perfect faces as close to twenty-five years, on average, the proportions of the lips (their fullness) was characteristic of fourteen-year-old faces, while the length of the face, from eyes to chin, was shortened to that typical of an eleven-year-old.[44]

Just as male peacocks have competed with each other to be very colorful, so women have competed to exaggerate the impression of youthfulness and health conveyed by their faces, as already noted. The reproductive significance of youthfulness for women helps to explain why they are prepared to spend a lot of money on cosmetics that create an impression of youth and health. Expensive creams promise to remove wrinkles, those telltale signs of aging, and to restore the healthy glow of youthful skin. Lipstick exaggerates the hue and enhances the fullness of the lips, making them seem younger. Shampooing and brushing make the hair seem more luxuriant and healthy (see fig. 4). Women use a whole range of products to enhance the apparent size and brightness of their eyes in an attempt to recreate the breathless and starry-eyed innocence of youth. What women do individually with makeup is analogous to what natural selection has been doing for hundreds of generations.

Humans are unusual in that both sexes evolved physically attractive traits through sexual selection. Human beards, for example, advertise biological quality in the same way that the gaudy plumage of the peacock does. This implies that both men and women compete among themselves for desirable mates. Handsome

Fig. 4. This beautiful young woman exaggerates her sexiness and health by using a corset to narrow the waist and brushing her hair to enhance the impression of healthiness (see text for explanation). (Reproduced from the Collections of the Library of Congress, LC-US-Z62-101143)

men and beautiful women are healthier and more fertile than their romantic competitors and transmit these qualities to children. Advertisement of biological quality through evolved bodily signals is not the only form that reproductive competition takes. People compete for marriage partners by being friendly and kind, and thereby advertising the qualities that are desirable in a long-term partnership. Men enhance their sexual attractiveness by competing for social status. This competition can take the form of reckless aggression, particularly in young men.

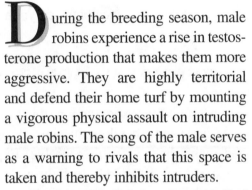

Chapter Three

Love's
Labors

Dating

Competition

and

Aggression

During the breeding season, male robins experience a rise in testosterone production that makes them more aggressive. They are highly territorial and defend their home turf by mounting a vigorous physical assault on intruding male robins. The song of the male serves as a warning to rivals that this space is taken and thereby inhibits intruders.

Singing, like fighting, is controlled by testosterone. This phenomenon has been demonstrated by injecting female songbirds with testosterone. Female songbirds normally do not sing, but they will do so after they have been treated with the male hormone.

The increased aggression of males during the breeding season is an adaptation that helps them succeed in the struggle for access to reproductive females. Their testosterone surge primes them to defend a territory by singing and fighting. The territory is of critical importance for reproductive success. Males that cannot defend a breeding territory do not acquire mates.

Human males do not have a breeding season, of course, but they compete most vigorously for mates when they are young, and when their testosterone levels are highest. Young men do not have breeding territories either, but they fight over something that plays the same role, namely, high social status among peers. Young men who have low social status—who are not "cool"—become the butt of jokes and are often viewed by girls as undesirable dating partners. Concern over "face," or status, is at the root of much apparently senseless violence between young men.

TESTOSTERONE AND AGGRESSION

Testosterone causes aggression: this matters to farmers who castrate male cattle and other domestic animals to reduce their aggressiveness and make them more manageable. Experimentally castrated male rats are much less likely to fight with other male rats. When they receive hormone replacement, their customary belligerence returns.

Of course, performing such experiments on human beings is not possible, although "natural experiments" have occurred. Since 1996 in California and a few other states, violent sex offenders get the option of chemical castration instead of confinement in prison. Parolees receive a weekly injection of Depo Provera (also a birth control drug used by women). Depo Provera lowers the blood testosterone level of treated men, which reduces the sexual fantasies typical of exhibitionists and child molesters when they commit sex crimes.

Research has shown that Depo Provera does reliably reduce sexual desire in men, although it has no impact on sexual desire in women and therefore cannot be used for female sex offenders. Chemical castration greatly reduces recidivism (reoffense) rates of paraphiliacs (e.g., child molesters, exhibitionists) from more than 90 percent recidivism to approximately 2 percent recidivism per year. This means that chemically castrated parolees present no greater threats than the males in the overall population.

Testosterone is also associated with aggression of a nonsexual nature. That is why so many violent crimes are committed by young men. The connection between age, sex, and aggression is so strong that if you wanted to solve the problem of violent crime in this country, all you would have to do would be to lock up all young men between the ages of thirteen and thirty—not that any sane person would advocate such an absurd measure. These are the peak years of testosterone production and of violent crime.

On December 1, 1997, teenager Michael Carneal reportedly opened fire on thirty-five classmates during a prayer meeting at Heath High School in Kentucky. His actions have been likened to a dream sequence from the movie *The Basketball Diaries*, although the fourteen-year-old himself emphatically dismissed this notion. He killed three girls and wounded five other classmates. Three years earlier, in the same town of Paducah, Kentucky, a sleepy community of 29,000 people on the banks of the Ohio River, teens reenacted a scene from the movie *Menace II Society*, which left one person dead and another wounded.[1]

There have been many other "copycat" crimes in which the perpetrators imitated life, art, or some mixture of the two. Commentators often puzzle over the common causes behind such tragic incidents and generally ignore the two most obvious and important factors, the actors and the implements of destruction. To be specific, the murderers were young, and they were male.

Young men are highly prone to physical aggression and recklessness. The violence-prone inclinations of young men can be enough of a problem even if they do not have access to lethal equipment, such as high-powered hunting rifles and assault weapons designed for use by soldiers in combat. When young men have ready access to extremely dangerous weapons, it magnifies the tragic consequences of their impulsive aggression.

Most social commentators who have asked why these tragic

episodes have occurred have focused on relatively superficial issues, such as where the killers got the idea for their actions or who had insulted them, thereby provoking the attack. These are trivial issues because young men who are interested in violence generally have no trouble coming up with a plan of action, pathetically misguided though it may be, and to the aggressive individual, there are many social situations that can be interpreted as insulting and calling for a violent response. What matters is not so much the precise script followed, but the way that the brains of young men resonate to them. Their aggressive behavior is so impulsive and so lacking in empathy for the victims that it can be thought of as originating in the older "reptilian" core of the brain that mammals inherited from their cold-blooded ancestors.

Psychiatrist Paul MacClean once commented that besides themselves, a client brings an anteater and a crocodile into therapy. He was referring to a three-part division of our brain structures (neocortex, limbic system, reptilian brain) corresponding respectively to rationality, emotion, and reflexes. The three brains are arranged in layers. An intriguing feature of brain anatomy is that the oldest (reptilian) brain structures lie at the base of the brain. The newest layer, the cortex (or "human" brain) lies at the top, and the limbic system (corresponding to the anteater, an ancient mammal) is sandwiched between the other two.[2]

Most people have experienced an episode of uncontrollable rage that is commonly called "losing it." The "it" that is lost is rational control by the neocortex. A useful tip in such situations is to take a deep breath and count to ten. By exercising the rational functions of language and counting that are organized in the cortex, control by the cortex can be restored. Young men are particularly aggressive because testosterone primes the subcortical structures mediating aggressive behavior. Why this pattern should have emerged amongst our ancestors has been the subject of intense scrutiny by evolutionary psychologists.

The Young Male Syndrome

The killing of members of the same species is surprisingly common among nonhuman mammals, including lions, elephant seals, and chimpanzees. This contradicts the view of early behavioral biologists that animals engaged only in "gentlemanly" aggression through which rivals tested each other. If the stakes are high enough, male mammals will fight to the death.

This phenomenon is clearly illustrated in the case of elephant seals during the breeding season. Females congregate in the territories of dominant males. Rival males compete vigorously over these reproductively valuable pieces of real estate and deliver vicious bites that are sometimes fatal. The bulls are often so belligerent that they attack and kill pups unlucky enough to cross their path.

Indiscriminate aggression is probably less common among humans. Thus, most observers are struck by the apparently unmotivated nature of most of the school shootings. The net benefit to the homicidal teens is the prospect of a momentary thrill, perhaps similar to the experience of watching a violent movie. The downside is life in jail, or even a death sentence. Any rational decision maker would draw the same conclusion: One should not open fire on classmates or the public! The costs far exceed any possible gain.

"Trivial altercation homicides" (to use legal jargon) are the most common type perpetrated by men in their teens and early twenties. The most common occasion for such killings is a dispute between young men before an audience of their mutual acquaintances—this helps explain why high school has been chosen so often as a venue. The conflict begins over some trivial pretext, such as who is stronger or who is next in line to play pool. It escalates when neither party is willing to back down and lose face.

Trivial altercation murders are not trivial at all, but mask competition over something important, namely, place in the pecking order, or social status. A recurring theme among recent high school shooters has been the perceived lack of respect from others and the desire to make a name for themselves. If there is anything rational, though twisted, about the conduct of these individuals, it involves the quest for prestige in the eyes of peers.

Evolution has ensured that young men are particularly concerned with their social standing because this increases their attractiveness to women and thereby enhances their mating success. Just as male robins are aggressive about defending their territory, so young men are hotheaded about their reputations. This is more true of young men than middle-aged ones because their status relative to peers is still being settled. An evolutionary history of mating competition between young men has produced an increase in aggressiveness during the teen years that coincides with and is mediated by elevated levels of testosterone in the blood.

The modern environment, replete with deadly instruments such as automobiles and guns, makes for dire results when these are in the hands of impetuous young men playing to an audience of peers. The appeal of aggression for young men can be thought of as a throwback to the evolutionary stage of reptiles because the neural machinery supporting it lies in the ancient part of the brain inherited from the Age of Reptiles. The reptilian brain is normally suppressed or inhibited by the neocortex, but this "balance of power" is shifted by elevated levels of testosterone in the blood that prime reptilian circuits of aggression.

SEX DIFFERENCES IN AGGRESSION AND RISK TAKING IN THE MODERN SETTING OF THE AUTOMOBILE

The greater aggression of young men and their willingness to take risks results in carnage on the roads. Women are much safer drivers because they are less vulnerable to highway games of one-upmanship. Canadian teenage male drivers are more than two-and-a-half times as likely to die in road accidents as women, even with adjustment for the greater mileage driven by men. The relative number of accidents by males declines steadily with age so that a man driving in his mid-seventies is almost as safe as a woman of the same age. Little solace can be taken from this fact, for by the time the average woman reaches her seventies, her driving has become very unsafe, according to accident statistics.[3]

Women are safer on the roads because they take fewer risks and therefore tend to stay clear of trouble. Evolutionary psychologists Martin Daly and Margo Wilson at McMaster University in Ontario concluded that the dangerous driving of young men is part of an evolved masculine tendency to take risks "that is often dysfunctional in evolutionarily novel settings such as automobiles."[4] It is essentially a display behavior analogous to chimpanzees making their hair stand on end to make themselves look bigger and crashing around in the jungle to impress their rivals.

Young men often take driving risks in situations where they can impress other young men with their daring and skill. This conclusion came from a formal study showing that men were much quicker to make a dangerous turn into traffic if they have another man aboard than if they are driving with a woman. Women's driving did not depend on the sex of the passenger.[5] When men drive with other men, their behavior gets riskier. This automatic shift in bravado in the company of peers would have helped our male ancestors to establish high social status. Young women are attracted to young men who have high social status, although they may not like the risky behavior through which it is obtained.

Daly and Wilson see dangerous driving as part of a pattern of reckless and aggressive behavior designated the "Young Male syndrome." Macho posturing in young men enhances their status in the eyes of peers. This would account for flashy but pointless lane changes and the other antics of young male drivers that contribute to senseless tragedies on the roads.

Competition over social status can take the form of complete strangers getting caught up in spontaneous racing and games of chicken. These dangerous games share the common element of an aggressive challenge between young men. "Road rage" is another example. One driver may accidentally cut off another and the incident escalates into a life-or-death struggle over face in which the "aggrieved" party attempts to run the competitor off the road. It takes two to tango, and road rage normally occurs only when the initial altercation has been followed by a vigorous exchange of nonverbal signals, such as giving the other person the finger, that escalate the dispute. Such combatants can make life on the roads a gladiatorial experience.[6]

The aggression and recklessness of young men are oversimplified by attributing them to the effects of a single chemical on the brain. Yet elevated aggression in young men *is* partly due to "testosterone poisoning." The most compelling evidence linking testosterone and aggression in humans is the correlation between testosterone and violent crime as a function of age and sex in all countries around the world.[7]

Large-scale study of testosterone levels in soldiers has shown that high testosterone levels in the blood predict a wide range of social problems. Soldiers with high testosterone are more likely to have a history of aggression, to use illegal drugs, and to go AWOL (absent without leave). Their romantic relationships are fraught with conflict, and they are less likely to marry. When they do marry, their marriages are conflictual and violent and more likely to end in divorce.[8]

Although testosterone plays a role in violence, high testosterone men do not *necessarily* have a history of aggressive confrontations. The male hormone seems to have more of a permissive than a direct causal role. In other words, men with very low testosterone levels tend to be timid and are unlikely to engage in physical confrontation. Men with high testosterone levels may behave aggressively, but whether they do is an issue of impulse control. Impulse control itself has a known chemical basis, being strongly affected by the chemical transmitter *serotonin*, which is underactive in the brains of violent criminals, for example.

To summarize the evidence, the case against testosterone is strong! The greater physical aggressiveness of men relative to women parallels the greater aggression of male mammals of many other species, but not all. Testosterone reaches its highest level at the age when more men commit violent crimes, i.e., between the teens and late twenties. The most violent criminals are those with higher testosterone levels. High testosterone is associated with problems conforming to social norms in institutions such as the army and marriage, and predicts problems in both.

Despite this evidence, which can hardly be contested, testosterone has had a bad rap because its beneficial characteristics have been downplayed. The "mindless beast" interpretation of testosterone's effect on human behavior is challenged by recent research that compared the self-reported thoughts of men with their testosterone levels. High testosterone men are actually more thoughtful than others and more likely to report that they are wrestling with a problem. They are usually highly energetic individuals who are both more intellectually busy and more socially active than most men. Far from being sex-obsessed monsters, they are more oriented toward peers than lovers and are more likely to report thinking about friends than about girlfriends.[9]

In addition to its complex psychological effects, testosterone also builds muscle, and the adolescent spurt in testosterone production is responsible for the development of greater muscle mass in men than in women. Yet the assumption that high testosterone men are all brawn and no brain is a false dichotomy. This point is illustrated by building up a profile of what we know about high testosterone men:

- They hate to follow rules and regulations.
- They are argumentative.
- Their marriages are troubled or unstable.
- They are devoted to their male friends.
- They are intellectually and physically vigorous.

This profile is perfectly descriptive of the life and character of many creative men. The most creative period in the lives of men also corresponds to their age of highest testosterone production. Male artistic endeavors as diverse as composing jazz music and writing romantic poetry peak in the late twenties and decline rapidly commensurate with testosterone levels during the thirties.[10] Testosterone changes the brains of young men in ways that make them more willing to take risks, buck convention, and assert themselves. These changes can favor artistic creativity and can yet promote aggression.

Acknowledging that there is a chemical trigger for extreme aggressiveness in humans is different from claiming that our genes code for violence. Testosterone levels are not determined by genetics in a simple way, and actually vary widely in the course of a day. Thus, among rodents, if two male mice fight, the testosterone level of the victor rises, but that of the defeated mouse falls. The same phenomenon has been noted for human athletes winning or losing a competition.

The levels of key brain chemicals affecting aggression, particularly the chemical serotonin, can also be affected by *experiences* as well as *genetics*. While the uninhibited aggression found in many criminals could be due to genetics, according to behavior geneticist David Lykken of the University of Minnesota, it could also be entirely due to abusive and neglectful parental behavior without a genetic history of impulsive violence. According to this view, brain serotonin levels can be low due to genetic factors, or they could be reduced by a stressful childhood—or both influences might be important.

Serotonin levels of the brain can be reduced following alcohol consumption, so it's entirely possible that the high crime rates observed near bars are no accident: intoxicated people have low brain serotonin levels and reduced impulse control, predisposing them to violence. Even diet might play a role. For example, rodents fed a low-tryptophan diet have low brain serotonin and are more aggressive. (Tryptophan is the protein from which serotonin is made.[11])

Hard biochemical evidence links low serotonin to violent impulsive behavior in humans. In one follow-up study of killers and arsonists after release from prison, not only were those with low serotonin levels more likely to commit another violent crime, but serotonin levels before release predicted recidivism or reoffense with 84 percent accuracy.[12]

SEX DIFFERENCES IN PHYSICAL AGGRESSION

Men are more physically aggressive than women at all ages and in all societies. This kind of uniformity has impressed evolutionary psychologists, and it underscores the folly of pretending that if you treat boys and girls the same, they are going to turn out the same.

Psychologists often argue that giving a doll to a little girl or giving a tool set to a little boy conveys important messages about the kind of skills they are expected to develop. While this is undoubtedly true, boys and girls differ in their inclinations no matter how they are treated. As we've seen, these different inclinations determine the choice of toy more than the choice of toy determines the inclinations. This point can be illustrated by the greater interest of most boys in games of aggression.[13]

Parents who inculcate nonviolence in their sons by keeping them away from violent toys and violent TV are often alarmed to discover that the sons use their imagination to turn common objects like sticks and pinecones into weapons of destruction, such as spears and hand grenades. Boys are intrinsically more interested than girls in boisterous and aggressive play. This sex difference is found in monkeys and other primates, and experiments reveal that it is mediated by early levels of testosterone. Female monkey infants exposed to testosterone in the womb (by injection of the mother) exhibit male-type boisterous play. This means that even before birth some basic differences in patterns of interest have been laid down in the primate brain. Women may be just as aggressive as men in the sense of being willing to start an argument, but they are generally less likely to resort to life-threatening aggression.[14]

Despite this pattern there is some overlap between the sexes, with some women being more physically aggressive than some men. Sex differences in aggressiveness also change according to context (women can be quite physically aggressive toward lovers and children, and women in highland New Guinea go to war). We nevertheless ignore the patterns of sex differences in aggression, at our peril. Thus, men commit about nine times as many violent crimes as women. They make up an even higher proportion of combatants in the world's wars.

PLAYING AT AGGRESSION

Although women have recently become much more active in sports than previously, the overwhelming cross-cultural pattern has been the monopolization of sporting activities by men. The current level of participation by women in sports is unusual but not unique. In the ancient Greek civilization of Sparta, physical fitness was emphasized in the education of women as well as men. Young people of both sexes exercised together in something akin to modern health clubs, the only noticeable difference being that the Spartans were probably completely naked, according to philosopher/historian Bertrand Russell.[15] The physical education of young women was designed to toughen their bodies so that they could nourish sturdy warrior sons in their wombs. To this end, they exercised by running, wrestling, and throwing the javelin.

The mystery of why men everywhere are much more interested in contact sports begins to make sense when one realizes that competitive sports arose primarily as a form of training for battle. Since making war is predominantly a masculine activity, it makes sense that men should be more interested in the training process. Sports and warfare both stimulate the brain structures implicated in physical aggression.

Austrian ethologist Konrad Lorenz was one of the first behavioral scientists to connect warfare with sports.[16] Lorenz felt that sports competitions between nations are substitutes for warfare and that they allow our aggressive drives to be harmlessly defused, a process known as *catharsis*, which was described by Sigmund Freud. This displacement theory of aggression has fundamental flaws, however. One is that expressing anger does not cause the emotion to go away and may actually strengthen it. A demonstration of this problem for Lorenz's theory comes from a study of California workers who received pink slips. After they had been fired from jobs in the aerospace industry, the workers were first asked to verbally describe their feelings about their former supervisors and employers. Then they expressed their feelings in writing. Instead of being cathartic, the interviews made the aerospace workers even angrier. Ruminating about their grievance did not reduce their anger, but actually increased it.[17]

Many team sports have the structure of simple warfare in which there are two sides, each with its own territory on the field and each aiming to invade the territory of the other to obtain a symbolic prize (e.g., scoring a goal). The basic objectives of attacking or defending territory are equivalent to those of conventional warfare.

The connection between sporting and military conflict is much more immediate in the minds of people from societies in which warfare is still commonly practiced. This point is underscored by the adoption of the sport of cricket among the Trobriand Islanders. As described in a classic ethnographic film, the islanders took that most staid and stylized of English team sports and turned it into a recognizable war game, complete with full battle dress, body paint, chanting, and war dances. In societies where warfare is a common experience, sporting events may be little more than practice for fighting. They may well serve the function of a dress rehearsal in which rival villages have an opportunity to impress each other by displays of strength, agility, and courage. They are the preindustrial equivalents of modern military exercises often carried out to impress, or even to threaten, neighbors.

Given that sports have their origins in warfare and that competitive sports express hostile impulses, their appeal—particularly to men—makes sense. They provide the thrills of fighting without having to face real dangers. Sports provide such a stylized expression of aggressive impulses that they can be played without the fear experienced of men going into battle. Just because people may enjoy the ritualized expression of aggressive impulses when they play sports does not mean that their aggression is defused, as the catharsis theory would predict.

The process of stylizing aggressive behavior has been taken one step further by the entertainment industry, which allows people to experience the vicarious thrill of expressing aggression without any fear whatever, and without any risk of retaliation. The process finds its ultimate expression in video games in which players themselves produce the actions that kill their virtual opponents. Video games thus provide an opportunity for the pure expression of aggressive impulses. They may also be used to reinforce the learning of practical military skills.

TV AND VIDEO VIOLENCE

The phenomena of virtual violence are often thought of as peculiar to the modern world and therefore largely irrelevant to evolutionary interpretations. Yet a rudimentary understanding of the adaptive nature of human aggression helps us to understand some of the most important features of virtual violence, such as its appeal to boys whose brains are primed for physical aggression. This aggression evidently played an important role in reproductive success in the evolutionary past. Just as an evolutionary perspective helps us to understand the "senseless" carnage of road rage, so it gives us some critical clues to the puzzle of recent high school shootings in the United States. The young male killers were engaged in a quest for higher social status, another example of the young male syndrome.

If allowed to watch TV unsupervised, many children develop a strong taste for viewing aggressive action movies. Such shows can influence the behavior of children in the superficial sense that children attempt to imitate many actions they have seen. For example, the popularity of martial arts in entertainment is reflected by the frequency with which karate chops and kicks are displayed in the normal interactions of boys in the company of their friends. Is this true aggression or mere fantasy? Does watching violent TV make children more prone to committing dangerous aggression as adults?

Researchers have clearly demonstrated the imitation of aggression by children in controlled laboratory experiments. Social psychologist Albert Bandura, working in the 1960s at Stanford University, found that if children were allowed to play with large inflatable ("Bobo") dolls after they had seen an adult strike the doll with a mallet, they would also use the mallet to strike the doll.[18] Whether this should be called aggression or play is controversial, although the involvement of the same brain structures may make this distinction unimportant. This point becomes particularly troublesome concerning highly realistic, aggressive video games.

Television is a source of many violent images for children. Its violence can cross over from simple imitation, which can be innocent in intent, to everyday behavior. In one study, preschoolers who viewed violent cartoons each day were rated as more aggressive toward other kids on the playground.[19] This phenomenon is familiar to parents whose children imitated the violent fantasy world of *Power Rangers*. Children liked some martial activities so much that they constantly imitated them in play, creating an impression of viciousness that alarmed many adults.

Children may attempt to imitate some aspects of violent depictions that they have witnessed, which can increase the aggressiveness of their play with other kids. The real question is whether these fantasy-related actions penetrate into real-world situations in later life, as in using guns to kill people.

Are people who have watched a great deal of violent television programs in childhood more likely to turn into violent adults with a greater likelihood of committing violent crimes? The answer, according to a 1972 study by social psychologist Leonard Eron of Yale University, is that they are. Eight-year-olds who preferred violent TV programming were rated as more aggressive by their peers at eighteen years old, and they scored higher on a delinquency questionnaire. Interpretation of such findings has remained controversial, however.[20]

The news media like to connect violent crimes with fiction; hence the connection between high school shooter Michael Carneal and *The Basketball Diaries*. It was true that he had seen the movie, but he had dismissed it as boring—although this reaction could be an adolescent pose. All such copycat scenarios are weak because they do not explain why, out of all the thousands of episodes of fictitious violence witnessed by teenagers, this particular one was selected—apart from the fact that it occurred in a high school. Moreover, it does not explain how a youth with no military training could engage in the cold-blooded execution of three young women who were not only acquaintances, but friends. This riddle can be partly solved in terms of the world of violent video games with which Carneal seems to have been fascinated.

According to psychologist David Grossman, currently at Arkansas State University and a retired army officer, there is a natural aversion to killing other people without provocation, which is quite a problem for military strategists. In World War II, for example, fewer than a fifth of the U.S. troops were estimated to have actually fired their gun in battle. For this reason, the military introduced man-shaped targets. Shooting games with these targets helped recruits overcome their aversion to killing the enemy. Video games have taken desensitization to violence to an extreme and also reward players for good shooting ability.[21]

This connection between video games and Michael Carneal's actions has led the families of the three victims to sue the company which makes some of the most violent games. According to the attorney for the families, Mike Breen, "Michael Carneal clipped off nine shots in about a 10-second period. Eight of those shots were hits. Three were head and neck shots and were kills. This is way beyond the military standard for expert sharpshooting. This was a kid who never fired a pistol in his life but because of his obsession with computer games had turned himself into an expert marksman." In fact, however, Carneal had the pistol in his possession for a couple of days before the shootings and had apparently engaged in some real-world target practice. Of course, Breen's account is only one side of the story and may not be the complete picture.[22]

The fascination that draws children, particularly males, to violent video games is probably built into their brains, along with their attraction to aggressive games and toys in general. Yet David Grossman argues that, just as pilots train on flight simulators, video games can be considered murder simulators. This is particularly true of arcade games that feature realistic pistol grips complete with recoil. To call some video games "murder simulators" is no longer a metaphorical claim because the Marine Corps is adapting a version of one of the most popular violent games to train its own personnel for combat situations.[23]

Video games not only train players in accurate marksmanship, but they also desensitize boys to the bad consequences of violence. Newer, more realistic games feature pleas for mercy, screams of agony, naturalistic wounds, and pools of blood. Desensitization to the consequences of violence is achieved partly through simple habituation but some games, such as *Doom*—in which the lone gunman confronts a variety of monsters—have a progressive structure in which the player is rewarded for kills by receiving more powerful and gorier weapons. The shotgun is replaced by an automatic weapon and eventually a chainsaw is used on the enemies.

Video games clearly contribute to the means of committing violent crimes, in the sense of providing training and desensitizing the shooters to gore. The actual weapons used are rarely difficult to find in a country in which there are more guns than people. This leaves unaddressed the vital question of motive where the evolutionary perspective provides its most valuable insights.

The study of precipitating events for homicidal violence among young men reveals that the most frequent context is known as a *trivial altercation.* It might seem that young men are losing their lives over something unimportant, such as who was first in line at a store. From an evolutionary perspective, the apparent cause of the dispute is really only a pretext. Young men get into scrapes not just because they are predisposed toward physical aggression, but because they are sensitive to situations that might adversely affect their standing in the eyes of peers. Trivial altercation homicides are fought over "face" or social status that is not a trivial matter at all. To allow an insult to pass unchallenged is to suffer a decline in social status and hence a loss of attractiveness as a dating partner.

The importance of prestige in the eyes of peers is a subtext that runs beneath the surface of most or all of the high school shootings. For example, Littleton, Colorado, killers Eric Harris and Dylan Klebold belonged to a clique known as the "Trench Coat Mafia," whose members were constantly ridiculed and insulted by their archenemies, the "jocks." In a school with a reputation for sports accomplishment, the jocks were treated like gods by teachers and administrators. Their high profile also made them attractive to female students.[24]

One way that the jocks purportedly asserted their superior social status was to insult Harris and Klebold and others at will. A friend of Harris's, Jeni La Plante, said, "The jocks would walk up to them and call them fags." If this is true, Harris and Kliebold might have decided to get even with the jocks—and everyone else—in this inappropriate, heinous way.[25]

There is a curious pattern to the rash of school shootings that extended from 1997 to 1999, with sporadic subsequent outbreaks. All of the perpetrators were young white males. All purportedly had trouble fitting in with their peers and were reportedly either subjected to insults or felt that they were not being taken seriously. They were ignored or despised by most of the people who knew them. Given their low social status, it is no surprise that most had trouble attracting girlfriends. Of course that does not excuse what they did.[26]

By any reckoning, Michael Carneal was very low on the totem pole at Heath High. A little more than five feet tall, weighing only

110 pounds, and the wearer of thick spectacles, he was far from physically impressive. His physical immaturity was matched by psychological childishness. For example, he still had a taste for the music of the Smurfs at a time when his peers had moved on to "gangsta" rap. As if this were not bad enough, he lived in a household in which his older sister, Kelly, a popular girl and star student, received much of the family attention. Hungry for notice, Carneal played the class clown, performing stunts such as stealing CDs, passing parsley off as marijuana, and distributing pornography in class.[27]

Carneal was not a big hit with the girls. Although he had begun to date, he had never so much as kissed a girl. Other students reportedly called him "faggot," and a school gossip sheet reported that he was gay. In the weeks before the shootings, he developed an unrequited passion for Nicole Hadley. When he opened fire on the prayer circle, she was his first victim.[28]

After he had stolen the guns that would be used in the school shooting, Carneal's first impulse was to bring them to the house of a friend, where the boys amused themselves by shooting at targets. He entertained the fantasy that having the guns would make him popular among other boys and said, at first, that he never dreamed of shooting anyone.

On Monday morning, he took the arsenal, wrapped in a blanket, out of his sister's car. He had a Ruger .22 pistol in his backpack. His friends were hanging out in the front lobby of the school. The morning prayer circle was winding down. Carneal told the others what was in the blanket and looked forward to the attention and notoriety the guns would bring. The boys examined the guns with polite interest but were not unduly impressed. After a couple of minutes, they had moved on to the next topic of conversation, new CDs. Frustrated, Carneal put his hand into the backpack and withdrew the Ruger. He put plugs in his ears and shoved the ammunition clip into the pistol. Thinking that these were the antics of the class clown, no one paid the slightest attention, until he began shooting.[29]

In his own words, "I don't know why I wasn't bluffing this

time. I guess it was because they ignored me. I had guns. I brought them to school, I showed them to them, and they were still ignoring me." Similarly, the urge to impress peers with reckless behavior manifests itself in dangerous driving and road rage as well. Although Carneal claimed not to understand his own motives, when he was interviewed in juvenile detention he bragged that "people respect me now."[30]

Chapter Four

The
Dance
of
the
Sexes

One of the most spectacular courtship displays of any bird is that of the great crested grebe. The courting pair dances right across the surface of the water with a synchrony of timing more perfect and breathtaking than anything achieved by Fred Astaire and Ginger Rogers. Grebes do not attend dancing school. The genetically programmed synchronous display is triggered in each member of the pair at the same instant, rather like two clockwork toys being released simultaneously. Dating among humans can resemble these grebes.

FEMININE WILES

The proverb "Faint heart never won fair lady" suggests that courtship is a heroic activity of men performing for fair ladies, who sit around waiting to be won, like stuffed toys at a county fair. However, a cool scientific appraisal of the antics of men and women in a singles' bar shows that it is the men who behave more like stuffed toys.

Austrian ethologist Karl Grammar, of the Ludwig Boltzmann Institute in Germany, observed the way in which romantic partnerships develop by unobtrusively recording the behavior of patrons in a nightclub in the late 1980s and subsequently interviewing them. The first puncture in the male heroism myth was delivered by the finding that men do not select their partners. In fact, men did not approach women without first being subtly invited to do so by nonverbal signals. Characteristic nonverbal invitations by women consisted of looking at the man, lowering the gaze and looking away, or simply gazing at him steadily for several seconds.[1]

We call such coy courtship initiations "flirting." There is a whole battery of flirtatious behaviors that is peculiar to women: high-spirited giggling; walking with a wiggle; seductive licking of the lips; batting the eyelids; bending over to accentuate the curves of the body; pressing a hand to the chest; preening the hair; and waving of the hands to indicate inoffensive helplessness.

Grammar's study found that once a man had been nonverbally invited, he would approach and strike up a conversation. This enabled the woman to assess his credentials at closer range. If he did not meet her criteria, he was quickly brushed off by a display of aloofness. A woman both proposes and disposes in the early stages of courtship. Grammar's work is confirmed by that of independent researcher Timothy Perper conducted in American singles' bars.[2]

If the woman is romantically interested in her prospect, she may unleash a barrage of nonverbal messages that few men can resist, according to the work of Grammar and Perper. She may stand close to the man, freely invading his personal space, and feel free to touch the man's shoulder, or his face. She sways her hips against his and may contact his back with her breasts "by accident."

Once couples become fully engaged in the courtship dance,

they develop a synchrony in their movements that is far less spectacular—but no less remarkable—than the synchronous dance of the crested grebe. Arm in arm, the couple indulges in prolonged gazing at each other. Their body language expresses mutual engrossment as they turn slowly to face each other and thus to exclude others. Their movements become synchronized. Couples studied in bars drank at precisely the same time. They even breathed in time. Dances are a formal social occasion where courtship may take place. Synchrony of movement evidently helps put couples in tune at an emotional level (see fig. 5).

Perhaps the most astonishing finding from Grammar's work was that the men thought they had been running the show—a striking testament to the strength and obtuseness of male ego! Female interviewees, in contrast, were coolly aware of what had actually happened. Women have an impressive arsenal of nonverbal weapons and many chose them consciously and deliberately. Our female ancestors were skilled at enticing men to become their partners. In other words, they secured the work and assets of a "husband" to support their children.

The courtship dance of the crested grebe may be as entertaining as any dance routine of Fred Astaire and Ginger Rodgers (see fig. 6), but grebes are not in the entertainment business. Evolution designed their courtship dance to satisfy two entirely different functions. First, just as dancing engenders romantic thoughts in human dating couples, so the courtship dance of the grebes gets them in the mood for mating. Second, it is a display of solidarity. It tells other grebes that these two go together. When a man and a woman become interested in each other at a singles' bar, their body language also says that they belong to each other while it subtly moves them along to the same emotional wavelength.

When describing the beautiful courtship of the crested grebe, it may seem inappropriate to point out that romantic relationships are full of potential for conflict between the sexes. Yet it is not out of place because the display itself is aimed at discouraging competitors.

Fig. 5. Synchrony of movement puts couples in tune emotionally. This couple danced at the San Angelo Fat Stock Show in Texas, in 1940. (Reproduced from the Collections of the Library of Congress)

Fig. 6. Motion picture poster of 1936 showing Fred Astaire and Ginger Rogers. (Reproduced from the Collections of the Library of Congress, LC-USZ62-101755)

It is saying, in effect, "If you are looking for a partner, look elsewhere because this one is taken!" Moreover, if acquiring a lifetime partner were a simple matter, there would be no need for any courtship rituals. These "dances" are part of the process through which two individuals persuade each other to "take the plunge" together. However, the need for persuasion arises from a conflict of interest.

DOMESTICATING MEN

The conflicting interests of men and women are as old as sexual reproduction itself. Going all the way back in evolutionary time to the first sexually reproducing animals on earth, females are the ones that produced the largest sex cells. Eggs are much more ener-

getically expensive to produce than sperms, which are tiny and produced in large numbers. Females thus invest more in offspring from
the very beginning. Males therefore compete among themselves for
access to the greater investment of the females.

While female investment in offspring is almost always very
large, male investment is highly variable. At a minimum, a male
might contribute only his sperm. At a maximum, he might take
responsibility for all care of the young. Surprising as it may seem,
the latter happens. For example, among jacana birds, it is the males
that incubate the eggs and take care of the hatchlings. Among most
monogamous birds, like grebes, there is a fairly equitable distribution of the work of caring for the chicks.

When both sexes cooperate to care for the young, this leads to
the establishment of very strong bonds of mutual affection that are
extraordinarily persistent. For grebes, swans, and human beings, the
pair bond can last a lifetime. Despite the strong bonds of love and
companionship, ancient conflicts of interest linger in the shadows,
one of the most important being conflict over sexual infidelity.

Male psychology evolved to become less choosy in the selection of a sexual partner. The cost to a man of being promiscuous
was not great, whereas for a woman a single sexual encounter could
result in the huge commitment of pregnancy and child rearing.
Women who mated indiscriminately would leave fewer surviving
offspring than women who were more cautious in the selection of
a partner. The more selective would choose men who were physically attractive, and thus had good genetic characteristics, or contributed to nurturing the child, or both. In either case children of the
discriminating woman would have a greater chance of surviving
whether this was due to superior genes, better nutrition, or better
protection from enemies and predators. This means that women
would have evolved a high degree of selectiveness in their choice
of sexual partners.

By contrast, the male partner in casual sexual encounters would
have increased his probability of leaving more children without

paying any of the costs of child rearing. Our male ancestors who enjoyed casual sexual relationships with several women would have been more reproductively successful than men who were faithful to a single partner. Their male children would have inherited the genetic basis for having a roving eye through the father's genes. This explains why modern men are more interested in casual sexual relationships than women are. The reproductive benefits of philandering exceed the reproductive costs for men, whereas the costs exceed the benefits for women. Needless to say, ancestral men were not necessarily interested in having babies, and may not even have made a connection between sexual intercourse and reproduction. Natural selection designed them to be interested in casual sex for its own sake and this tended to increase their reproductive success. The same argument can be made about the greater eagerness to mate of males of other species.

Like males of all species, men compete for the reproductive resources of women. At the same time, women compete over help from men in raising their children. Given that the sexes have evolved differences in their needs and objectives, there is a potential for conflict. Anthropologists see the marriage contract, in different societies around the world, as one solution to this conflict. If men agree to support the children of the marriage, in return they are given an opportunity to father them (see fig. 7).

Fathers in most societies perform some childcare services, but women everywhere do the lion's share of this work. Paternal support revolves around helping to provide food, shelter, and protection, each of which would have been critical for survival in the evolutionary past. In one of the few remaining subsistence societies, the forest-dwelling Ache of Paraguay, support of fathers is crucial. In fact, 45 percent of children who lost their fathers to death or divorce at any point before the age of fifteen died before themselves reaching the age of fifteen years, compared to a 20 percent mortality rate for children who had fathers present in their families. The absence of a father triples the risk of death due to illness, and

Fig. 7. A Bit of Romance, painted by Howard Pyle (ca. 1896), depicts love as the combination of sexuality and domesticity that defines the marriage contract around the world. (Reproduced from the Collections of the Library of Congress, LC-USZ62-048124)

doubles the risk of being killed by other Ache. Reading between the lines, one can see that if Ache women did not marry, their chances of raising children to maturity alone would be fairly bleak.[3]

Given that it is difficult to raise children, that a mother's very great investment in her offspring is lost if the children die before reaching maturity, and that the presence of the father provides considerable insurance against the hostile forces of nature, the case for bonding with a particular man and obtaining masculine support must have been compelling for our female ancestors.

Whereas women have no choice but to make a very great investment in each of their children, if only because they must carry the fetus during pregnancy, men can theoretically pursue different reproductive strategies. If they abandoned their children, far more would have died before reaching maturity. Yet they could have made up for high infant mortality by siring many children through many women. Instead of being *dads*, they could behave like *cads*.

The behavior of modern men reveals a certain amount of emotional ambiguity about whether to be cads or dads. (Of course women may also feel ambiguous about marriage, but this usually seems to take the form of asking whether their current partner is worthy of them rather than dreams of sexual gratification with a limitless supply of attractive men; female fantasies are more likely to involve being wanted by men.) Some men who are happily married and love their wives are perfectly capable of having extramarital affairs without feeling much guilt or remorse, at least while they are getting away with it. Yet there are only a handful of preindustrial societies in which children grow up apart from their fathers and in some of these the absence of their father is temporary. From the point of view of the evolutionary dance between the sexes, men have been pulled away from their cad tendencies and turned into dads. Instead of constantly wandering around trying to impregnate women, ancestral men at some point began to stay at home and take responsibility for their children. They became domesticated, in the sense of organizing their hunting activities around a home base.[4]

The notion of women domesticating men is slightly misleading, however. Men stayed at home and behaved like dads presumably because this maximized the number of children they could raise to maturity. Why could they not leave more offspring by behaving like cads all of the time? One possible reason is that the male relatives of their female love interest protected her, which would have made being a cad quite risky. Another plausible reason is that impregnating a woman takes a surprising amount of time. Whatever the reason, there is no need to see the sexual behavior of men as driven by conscious reproductive intent. Other species manage fine without any such intentions and so, in all probability, did our ancestors.

Modern couples deciding to have a child often experience several months, or even years, of steady sexual intercourse before the woman conceives. Unlike most other mammals, including the primate order to which we belong, women do not have a distinct period of sexual heat, or estrus, which induces mating at a time when impregnation is most likely. There is a very slight increase in women's sexual motivation around the time of ovulation, but the probability of having sexual intercourse is roughly the same at all stages of the menstrual cycle.[5]

Among humans, ovulation is cryptic, or concealed. This phenomenon could reflect the evolutionary dance of the sexes. According to University of Michigan biologists Richard Alexander and Katherine Noonan, concealed ovulation is a means by which our female ancestors got fathers to invest in their children. If men cannot detect when women are ovulating, then a cad strategy may not produce as many surviving children as a dad strategy. If one assumes that Alexander and Noonan are correct, concealment of ovulation is an effective technique by which women have reined in men's polygynous tendencies, inducing them to stay at home with their wives, fathering most or all of a woman's children and helping them to survive to maturity.[6]

Stated from the woman's perspective, women who had a very high sex drive around the time of ovulation, and who clearly advertised their reproductive condition to strange men, as female chimpanzees do through conspicuous sexual swellings, for example, would have been vulnerable to cads. They would have received less paternal investment for their children and would have raised fewer children to maturity. Concealed ovulation undermines a cad strategy and promotes a dad strategy because a man cannot time intercourse to coincide with a woman's time of highest fertility. For this reason he may be more reproductively successful by staying with a single woman and siring all of her children, rather than pursuing many different women with a low probability of impregnating any of them. Once again, these "strategies" are unconscious products of natural selection.

A long evolutionary history of living with men who were physically stronger likely promoted the more finely tuned interpersonal skills of women. Women made up for a relative lack of bodily strength by developing aptitudes for predicting and controlling the behavior of men.

WOMEN'S SOCIAL SKILLS

Grammar's study of modern courtship showed that women are socially more astute than men. They orchestrate social interactions so skillfully that they can control their date, even when the man believes he is running the show. The view that women are more verbal than men is more than a stereotype. Women score higher on verbal tests, speak more words in a day, are quicker to verbal aggression, are more articulate, get verbal responses out rapidly, have more friends, and spend longer amounts of time speaking to them on the telephone. Moreover, when women talk, they reveal more intimate and meaningful information about themselves. Women are better listeners. They tend not to let their attention wander in the middle of a long story. They are more willing to offer comfort to another person—a sex difference in empathy that is present even in young children.[7]

Women are more skilled in reading and using body language. Countless laboratory experiments have showed that they are more skilled at reading facial expressions and detecting nonverbal signs of lying, for example. This research backs up the findings of field studies on courtship interactions.[8]

Psychologists often point to different childhood influences in order to explain why women have better interpersonal skills. They argue that giving a doll to a little girl and giving a tool set to a little boy conveys important messages about the kind of skills each needs to develop. While this may be true, it is also true that boys and girls differ in their inclinations regardless of how they are treated, a point already made for the development of aggression in boys.[9] Parents who strive to inculcate nonviolence in their sons by keeping them away from violent toys and violent TV are often alarmed to discover that the boys imaginatively turn common objects into weapons of destruction. Sticks are guns or spears. Pine cones are hand grenades. Sex differences in aggression are largely due to biology, as already pointed out, but upbringing does accentuate them, as happens in warlike societies. Sex differences in social skills may also reflect evolved differences.

Thus, women's abilities to entice and manipulate men in a dating context would have helped to ensure male support for their children. In the past, they may not have had much personal interest in striving for political power and social status, but they were attracted to men who had these qualities and therefore acquired them by association. In other words, a woman who succeeded in marrying a high-ranking man acquired high social status for herself. Even today, when women's earning power immediately after college is almost equivalent to that of men, they still express the same emotional needs that helped them to obtain paternal investment in the evolutionary past.

THE EVOLVED ROMANTIC NEEDS OF WOMEN

Women are far more interested than men in the social status of potential romantic partners. This is obvious from many different lines of evidence. In personals columns, female advertisers often state that they are looking for a professional or for someone with a college education. Men specify this less. When asked what they want in a dating partner, women articulate the same desire. Young women often complain about the scarcity of men, despite the obvious fact that most live within a few miles of thousands or even millions of single men! The problem is clearly not in locating individuals with Y-chromosomes but in finding men who are of appropriate social status and who are willing to make a commitment to them.

Although women focus on meaningful emotional relationships with men, their choice of a marriage partner can be surprisingly pragmatic. At a high-school reunion, you notice that the class nerd, who had been rejected in school by all the girls at the time, shows up with a stunningly beautiful wife. Your mouth gapes. How could she have done it? Can't she see he's a geek?

As for the women they marry, the geeks and the meek often inherit the earth. They work hard in school instead of pursuing a social life. They get good grades, go to college, and enter safe professions. By contrast, some popular physically attractive young men get distracted from their studies, obtain mediocre grades, and rely on their personal qualities to pull them through college and job interviews. Geeks save money and acquire property. Women who dismissed the class nerd begin to change their minds as soon as they discern his economic success. Material success is an important criterion in mate selection for women because of an evolutionary past in which men who were good providers promoted the health and survival of children. Even obtaining enough food would have been a challenge for the single mother of young children who might have been simultaneously pregnant and breast-feeding.

Wealth alone is rarely enough to make a woman feel satisfied with her date or mate. She also needs to feel emotionally involved with her partner. That means that the man must be interesting and exciting to be around and attentive to her emotional needs. As John Townsend points out in *What Women Want—What Men Want*, it is this complexity in female desires that often makes their behavior puzzling:

> Most women want men who are loving and tender with them but they also want men who are winners. This desire for both types of investment often causes them to behave in ways that men find baffling. For example, women want more cuddling and tender shows of affection than men do, and women's complaints about men in this respect are perennial. But at times many women also want their male partners to act passionately, lustfully, and masterfully. A woman can alternate between wanting to cuddle and nurture the man, wanting the man to cuddle and nurture her, and wanting him to act as though he were ravishing her. And all of these alternations can occur in one encounter.[10]

She wants to be wanted, but she also needs the kind of consideration that ensures she can control her own fertility.

Women's complex emotional needs can be interpreted as mechanisms evolved to obtain spouses who are willing and capable of high investment in marriage—in both the economic sense and the emotional sense. These evolved needs change remarkably little with the altered social and economic status of women in the modern world of occupational near equality of the sexes. This point is borne out in the life stories of some of the women that Townsend recently interviewed. For example, "Gloria," a thirty-six-year-old businesswoman, had been married to one successful man before moving out on him because he fooled around with other women and neglected her. She says:

> I love powerful, aggressive, self-made men. They are exciting because they are winners. They are the gladiators of today. They have a kind of sexual energy that most men lack. But I know how

they are. They are selfish and self-centered and are not good lovers. Some of them are talented at the mechanics of lovemaking, but they are not being emotional with you, and they aren't into saying, "Hey this is your evening and I will just totally please you." They are into getting pleased and having their needs satisfied. They are used to women waiting on them—at work and at home.[11]

Despairing of happiness with the kind of successful men to whom she was attracted, Gloria experimented with dating two working-class men who provided more affection than her husband had. Although she found these relationships satisfying to a certain degree, she would not have dreamed of marrying either of her lovers. She simply had no respect for them.

Men who are ambitious, hardworking, and rich are highly desirable as husbands. As Gloria makes clear, men who are successful and powerful are sexually attractive. She would agree with Henry Kissinger's remark that "power is the ultimate aphrodisiac."

The man who had seemed nerdy in high school may have a number of advantages as a husband. He is hardworking, successful, and dependable. He is devoted to his wife and children. The chances that he will pursue other women are low. His lack of personal charisma also reduces the chances that other women will be romantically interested in him.

Here we get to the heart of the matter. If other women are not romantically interested in a husband, why should the wife be? One important reason may be that when people spend a lot of time together, including the sharing of marital intimacy, they become strongly attached to each other. Shared interests and responsibilities may also be important. However, the geek's relative lack of sex appeal makes him exceptionally vulnerable to the wife's marital infidelity. A woman who receives his parental investment, while allowing a more appealing man, with superior biological qualities reflecting superior genotype to father her children, is getting the best of both worlds so far as the quality of her offspring is concerned. Yet most women are not interested in such duplicity. They want all of the criteria for a desirable husband to be contained in the same individual.

The problem of going for the now-successful high-school geek was skillfully captured by Gustave Flaubert in his novel *Madame Bovary*. The heroine is married to a country doctor who treats her with slavish adoration. As the plot develops, she casts aside this security for the transitory bliss of an adulterous affair. Flaubert, a master of pathos, allows us to feel sympathy for the cuckolded husband at the same time that we sympathize with the sense of romantic suffocation of the heroine. Our romantic psychology was designed by natural selection to promote reproductive success, not happiness. That is why love can seem cruel.

Having a comfortable marriage is not equivalent to being happy. Emma Bovary had the same weakness for a sexually attractive partner that men do. Women are normally more successful at controlling such impulses for a variety of reasons both social and physiological. In particular, they have most of the social responsibility for preventing unwise sexual relationships because of a double standard that has historically allowed men to sow their wild oats. Emma Bovary was depicted as weak and childlike because she did not succeed in mastering her sexual impulses.

Whereas women have three essential needs in a husband—social success, physical attractiveness, and emotional support (indicating willingness to invest in the woman and her children)—men require only two of these in a spouse. A man wants a wife who is physically attractive and who feels and expresses love for him. Men are not strongly attracted to women who have higher social status than they do. If anything, highly successful woman today find it more difficult to get married than their less successful sisters because the need to find men equal to or wealthier than themselves considerably reduces the pool of potential partners they are willing to consider.[12]

Townsend's study of love and commitment among American couples illustrates the greater role of wealth in the marital desirability of men than women. This difference is found for other societies around the world. He uses the story of a man named Fred as a case in point.[13] This evolved psychological difference emerges loud and clear despite change in the economic role of women. Fred was a thirty-eight-year-old divorcee who worked as a financial analyst and had an annual income of $90,000. His wife was very physically attractive and she agreed to take a low-profile job while providing domestic and emotional support for Fred in his career. Moreover, she was a good housekeeper and cooked all of the meals. The problem was that she found herself stuck in a job she hated and began to reevaluate her situation. She considered going back to graduate school and getting a Ph.D. in literature. Mindful of the bad job situation in this field, Fred suggested that she go to medical school instead.[14]

Once she went to medical school, the relationship changed. Fred found himself doing more and more of the shopping, cooking, and cleaning, to the point that he felt his career was being damaged. His wife turned out to be a brilliant student despite her low self-esteem. She zipped through medical school and soon found herself in a practice in which she was earning twice as much as her husband.

About a year after she joined the practice, Fred received a warning from one of her partners, with whom he was having a beer: "If I was you, I'd watch out for Alister."[15] Alister was another of the partners in the practice. Fred was not really concerned about Alister at the time. His wife assured him that people were talking just because they happened to be good friends. Further, Fred was not concerned about Alister because he had so little respect for the younger man. Alister was widely known to avoid certain medical situations because he was such a poor doctor. He found it hard to believe that his wife could be romantically interested in such an incompetent. This was a bad mistake because his rival turned out to be extremely rich due to inherited wealth.

The fat really hit the fire when Fred had to leave his position for a year on a consultancy job. He assumed that they both might occasionally have extramarital sex during this time, but that it would not threaten what he believed to be a strong marriage. When he returned on weekends, he began to notice signs that Alister had been living in the house. Moreover, their sex life was really dismal. His wife was just going through the motions; she obviously hated what she was doing. Before long, they were divorced.

This relationship illustrates two differences in the sexual psychology of men and women. The first is that Fred's helping his wife get through medical school did not strengthen the marriage, as he had anticipated. Women are attracted to men whose social status is higher than their own. Since Fred's wife was now earning much more than he ever could, she was ready to move on to the next echelon—old money. This cruel mindset is not very unusual. Of the white-collar women with working-class husbands interviewed by Townsend, most were thinking of trading up. Just as women are motivated to satisfy their evolved need for social status in a husband, so men trade up on the basis of physical attractiveness. Exactly the same callousness emerges in men when they divorce their wives of twenty years to marry much younger, more physically attractive women. The only difference is that when men and women trade up in spouses, they generally use different criteria. Fred's wife indulged her evolved need for a husband of higher social status after she became dissatisfied with her present marriage. She allowed herself to be swayed by the sex appeal of a man of higher social class.

The other interesting sex difference is that Fred was willing to indulge in extramarital sex and did not see it as having any bearing on his marriage. On the other hand, once his wife had fallen for Alister's boyish charm, she had extreme difficulty in having sex with Fred. For her, as for most other women, sexual intercourse is reserved as an expression of love. Many of Townsend's male informants describe themselves as continuing to have sex with girlfriends for months even though they did not love the women, and in some cases, actively disliked them; this suggests that the male brain is designed to enjoy "casual," uncommitted sex for its own sake. The female brain, for the most part, is not.

This is a bold statement and one that seems to fly in the face of some modern evidence. Among Townsend's informants, for example, there were some highly sexually active young women who reported casual encounters with dozens of different partners. These women were not typical of the predominantly medical student population studied by Townsend, but they were not very unusual either. On the surface their behavior was indistinguishable from men, which appears to contradict the view that the romantic interests of men and women were designed differently by evolution through natural selection. Townsend's interviews clearly demonstrated that their motivations and emotional responses were very different from those of their sexual partners, however.

Unlike highly sexually active men, women who had many brief sexual relationships did not engage in them for the purpose of obtaining sheer physical gratification. Townsend's student interviewees almost always reported some ulterior motive for their one-night stands. One was that desirable men were objects of competition between friends. Some men were so attractive that seducing them was considered a challenge and worthwhile in itself. Success in this game reassured a woman of her own sexual attractiveness. Such casual relationships by women may be an artifact of living in an era with effective birth control techniques, in which social sanctions against premarital intercourse are weak. They might also be due to the extreme competition between female medical students over male medical students. Most of the men in their circle were more interested in cultivating physically attractive women who were not medical students.

Another surprisingly common motive for brief sexual relationships was revenge. A woman who felt betrayed by her boyfriend might get back at him by seducing his best friend.

Perhaps the most telling aspect of the sex difference in sexual psychology was the reaction of women to their own casual affairs. One woman, called Chantal, reported a lot of variability in her reactions. Sometimes she would think of her behavior as though she were a man, that is, "have sex without strings attached." On other occasions she would think of it more as a woman does, getting very emotionally involved with her sexual partner, worrying what he thought about her and if he would call her the next day.

Ultimately, Chantal reached the conclusion that casual sex was not for her:

> I used to think I could sleep around with whoever [*sic*] I wanted but now I find this demeaning. Sex is like giving part of yourself to a guy, and if you do this without any emotional involvement, you are showing a lack of respect for yourself. Now if I try to do that, I begin to feel used. My attitude changed when I grew up and found a serious boyfriend. I'm not sleeping around now, and I have a lot more respect for myself.[16]

A similar sex difference in motivation emerges in the study of *reasons* for having extramarital affairs. Dissatisfaction with a current spouse and the need for love are common motives for marital infidelity of women. Men who have extramarital affairs are just as likely to describe themselves as happily married as men who remain sexually faithful to their wives. Their primary motive is physical pleasure unencumbered by emotional baggage.

Women who have brief sexual affairs are very much the exception that proves the rule because, even though their behavior seems the same as that of men, their emotional reactions are quite different. Many women engage in casual relationships at a time when they are establishing their own desirability as dates. After they have seen what's out there, they become more circumspect and, like Chantal, pursue a serious relationship in which they can expect a long-term commitment from their partner. Even when they are most open to sexual variety, women do not persist in relationships that they perceive as having little future.

By contrast, some of Townsend's male informants admitted that they continued to have sex with partners they disliked for several months just because their physical needs were being satisfied. By contrast, some of the female informants revealed that they could not continue having sex with men in relationships that did not lead anywhere because they found themselves getting emotionally involved with the partner and realized that the longer the relationship lasted, the more difficult it would be for them to break up.

Townsend's conclusions do not stand alone, but are part of a body of converging evidence showing that men and women approach sexual intimacy differently. Thus, the greater male propensity for sexual gratification without emotional involvement provides the basis for the sex industries of prostitution, pornography, and phone sex, which cater to a clientele that is almost entirely male.[17] Men and women also differ in their motivations for marital infidelity, as we will see shortly. Women who are unhappy with their marriages are more likely to initiate an affair, whereas men who commit adultery are just as happy with their marriages as men who do not.[18]

What makes the between-sex difference in the emotional response to casual sex so compelling in Townsend's study is that the women involved were not simply following a traditional sex-role stereotype. Most of them were highly ambitious and successful, and participated in masculine team sports. They rejected traditional feminine roles and had no ethical problem with sex outside marriage. They believed that they could enjoy casual sex without any emotional commitment just as much as any man. They were, therefore, puzzled as well as disturbed by the intensity of their own negative emotional responses to sex without commitment.

SEXUAL INTERCOURSE AND PLEASURE

The courtship dance of the sexes is a complex interaction, and its natural denouement is sexual intercourse. Whatever the social complications that a man and a woman negotiate before winding up in bed together, it is clear that their motivation for engaging in sexual behavior is the pleasure that it yields both partners, which is usually intense. There seems to be little difference in the physiological mechanisms promoting sexual arousal and orgasm in men and women, except those that are determined by anatomical differences in the sexual organs and the lower reliability of the female climax.

The scientific study of sexual pleasure has encountered two real impediments. One is that sexual behavior is considered essentially private and thus unavailable for scientific scrutiny. The other is that scientific psychologists have always had a problem in talking about pleasure as something unobservable because it is subjective, or contained inside the individual, and thus supposedly unavailable for scientific observation. The answer to both problems has turned out to be surprisingly simple: if you want to reach a scientific understanding of sexual pleasure, then you should collect data on people's sexual behavior. This project was postponed for a long time because of the Western religious view that human sexuality was sinful, and that sexual impulses should be suppressed as much as possible. The work of the first serious scientific investigator of human sexual behavior, English physician Havelock Ellis, is instructive in this respect. His seven-volume *Studies in the Psychology of Sex* (1897–1928) was banned on charges of obscenity.

American sex researcher Alfred Kinsey (1894–1956) also encountered controversy, but succeeded in establishing scientific respectability of research in this area through numerous interviews asking people about their sexual practices and experiences. Kinsey's work suggested that various sexual practices like masturbation, homosexuality, and extramarital affairs were more common than many people wanted to believe. His work was flawed by poor sampling methodology, however, and he may have been more of a crusader for sexual liberation than the objective scientist many thought him to be. According to biographer James H. Jones, Kinsey, although married, was primarily a homosexual with unusual sexual tastes like masochism, and was biased toward concluding that such interests were more common than they really are.[19]

Gynecologist William Masters (1915–2001) and psychologist Virginia Johnson, both at the Reproductive Biology Research Foundation in St. Louis, did more than anyone else to expose the mysteries of sexual pleasure to scientific scrutiny. In doing so, they exploited a window of opportunity presented by the sexual revolution of the 1960s. They published *Human Sexual Response*[20] in 1966, a work that summarized a decade of scientific observation of human sexual physiology. Masters and Johnson married in 1976 but divorced in 1991. Between them, they trained thousands of sex therapists based on their physiological research.

Their scientific work focused on detailed objective descriptions of the physiological reactions of participants to stimulation of the genitals through masturbation, natural intercourse, and other situations, including a mechanical copulating machine that recorded responses within the vagina. This included a penis made of transparent plastic that allowed for continuous optical monitoring of changes in blood flow, lubrication, and lengthening of the vagina, among other measures. The artificial penis was moved using electricity and a woman could control the speed of thrusting and depth of penetration, both of which tended to increase as orgasm approached.

Needless to say, this was a very odd scenario for conducting scientific research, particularly when you consider that participants were not only hooked up to sophisticated physiological recording devices but were also surrounded by a team of human observers. Participants were volunteers who were paid for their time. Some were strongly motivated by the need for sexual pleasure, but most claimed that they participated in order to increase our knowledge of the human sexual response. Participants were guaranteed anonymity and went through a very careful orientation process that gradually exposed them to the unusual situation without threatening their comfort level. No ethical or legal problems ever surfaced from the research, indicating that participants were treated with proper respect and consideration.

Masters and Johnson made many fundamental discoveries about the physiology of the human sexual response and drew some important practical conclusions. Perhaps the most striking finding from their physiological recordings was that sexual response takes place throughout the entire body. When a woman is sexually aroused, for example, she breathes deeply. A sex flush spreads from the breasts to the abdomen, thighs, and buttocks. The heartbeat speeds up and blood pressure rises. The nipples become erect and the size of the breasts increases, in addition to many complex changes in the genital area. The sexual response of men is also a whole-body phenomenon having many similarities to the response of women, even extending to nipple erection for 60 percent of men.

The sexual arousability of virtually the entire human body helps to explain the great variety of sexual behaviors that precede intercourse, such as hugging, kissing, and stroking the skin. Virtually any form of tactile stimulation performed in a romantic context can increase mutual sexual arousal. Some people enjoy having their ears nibbled, for example, and others like having their backs scratched. For many people, foreplay is almost as enjoyable as direct stimulation of the genitals through intercourse or other means.

Masters and Johnson concluded that a woman's sexual satisfaction is unrelated to the size of the man's penis. They found that all women are physiologically capable of orgasm when appropriately stimulated. Perhaps the most important finding was that the overall pattern of sexual response is very similar in men and women, with a stage of excitement followed by a plateau of intense arousal and pleasure that leads to orgasm and is followed by a stage of resolution, or calmness, when bodily arousal decreases to normal levels. The main difference found in the research was that some women are capable of having many orgasms in rapid succession, whereas men are not. The fact that men and women obtain intense pleasure from sexual intercourse is clearly an important ingredient in their romance that helps them maintain a positive attitude to each other. Frequent intercourse also increases the likelihood of a woman becoming pregnant, and that helps to explain why sex is pleasurable from an evolutionary perspective. Thus, American women who had intercourse less than twice a week while attempting to become pregnant took an average of eleven months to conceive compared to seven months for women who had sex at least twice per week.[21]

The pleasure of sexual intercourse can keep people in a relationship for some time, as we have seen in the case of the male medical students; still it is rarely if ever the sole basis for a satisfactory long-term relationship, which requires emotional intimacy and mutual caring. Marriages can even persist in societies where couples go for long periods without having sex. This is most obviously true in societies where sexual relations are forbidden for several months following the birth of a child. Known as the postpartum taboo, it is very common for preindustrial societies where it helps to space out births. In fifty-eight such societies the taboo on intercourse lasted for more than six months, but only three of these societies were monogamous. In polygynous societies where a man may have another wife to satisfy his sexual needs, the taboo lasts for longer. Societies with a postpartum taboo do not suffer marital instability as a result. Thus, frequent intercourse is clearly not necessary for marriages to persist.[22]

Masters and Johnson emphasized the key role of the clitoris in women's sexual pleasure and pointed out that it is the only organ whose exclusive function is pleasure. Although orgasm is possible in the absence of clitoral stimulation, the clitoris usually plays a role in orgasm whether self-induced or produced by a partner. The mere existence of these physiological mechanisms confirms that sexual pleasure must have been important for our female ancestors, motivating them to have sex regularly. The point of view that explains away both the clitoris and female orgasm as merely side effects of developmental mechanisms producing the penis and male orgasm misses the practical importance of pleasure in women's sexual behavior.

Pleasure may not be the sole function of the female orgasm, however. English biologists Robin Baker and Mark Bellis, of the University of Manchester, have demonstrated that a woman retains more sperm in her reproductive tract if she reaches orgasm during, or shortly after, her partner's ejaculation. They propose, controversially, that this is one mechanism through which women increase their chances of being impregnated by a man whom they find sexually exciting.[23]

Sexual experience is based on physiological reactions that are largely shared by men and women, but it involves much more than physiology. Psychology matters a great deal also. It is not just a matter of having various parts of the body rubbed in the right way. How sexual stimulation is perceived is just as important as the stimulation itself. Thus, women are just as responsive to pornography, in terms of their physiology, as men are but rarely purchase pornography because they have little interest in it (see next chapter). Evolutionary psychologists interpret this in terms of evolved differences in sexual motivation. They also point to a generally greater interest of men in sexual experience and sexual variety, for their own sake.

Chapter Five

The Cheating Hearts of Birds and Humans

In the early 1970s, a female telephone executive was participating in a two-week training course that was being run in the Chicago hotel where she was staying. During the evenings, participants would congregate in a hospitality suite for drinks and socialization. Time began to drag. A game of bridge was proposed. The trouble was that there were only three bridge players. With some egging on by a male trainee, who was only a beginner himself but who offered to help her, the woman agreed to take a hand making up the necessary fourth player.

The card game was so much fun that it went on until three o'clock in the morning. The next day the same thing happened. The only problem was that the woman's husband had been trying to call her. Each night he stayed up until 5:00 A.M. periodically dialing the number of her room, but never receiving an answer. Finally, on the next day she answered the telephone in the afternoon and had to deal with a very angry spouse. The irate husband demanded to know where she had been when he called. She told him truthfully that she had stayed up late playing bridge with the other trainees. This made the husband even more upset because he knew that the wife did not even know how to play bridge and had never been interested in playing cards before.

Human marriage is a largely monogamous arrangement, but it is exposed to many challenges. Animal behaviorists have recently discovered that many of the bird species which were supposed to live in a state of happy monogamy are actually faced with an ever present threat of cheating. Take the case of the dunnock, a plain-looking bird that inhabits British hedgerows. A nesting pair is often accompanied by a satellite male which sneaks copulations in spite of the resident male. When the resident discovers this infidelity, which he apparently cannot stop, he pecks repeatedly at the repro-ductive tract of the female, forcing her to eject the rival male's semen.[1] Our hunter-gatherer ancestors were affected by mating competition also, judging from research on human semen produc-tion. This backdrop is likely responsible for the existence of intense sexual jealousy in modern men.

When Julia Roberts's character was stalked by Patrick Bergin's in *Sleeping with the Enemy*, this was just a pretext for the creepy horrors to follow. It seemed in the mainstream of thriller fiction. The plot is, unfortunately, lived out by hundreds of thousands of American women. Spousal battery is the leading cause of death for women between the ages of fifteen and thirty-five. This tragic fact is an extreme manifestation of the evolved tendency for jealous rage that would have helped our male ancestors to avoid being the victim of spousal infidelity. Before broaching this issue, it is helpful to know something about the evolution of the reproductive system of humans compared to other species.

ARE HUMANS MONOGAMOUS?

Swans are monogamous. Pairs mate for life and raise young together year after year. Most other bird species are monogamous for a breeding season. Among mammals, monogamy is rare, and reproduction is the result of open competition between males for access to females in heat. Among our closer biological relatives, the primates, all kinds of reproductive systems can be seen—from monogamy in gibbons, to polyandry in lion tamarins, harems in baboons and gorillas, promiscuous mating among chimpanzees—but monogamy is very rare. Apart from New World monkeys, and some lemurs and lorises, it is found only among siamangs and gibbons.[2]

Biologists classify humans as *facultatively* monogamous, which means that they tend to practice monogamy in cases where it is difficult for a woman to raise children without the support of a husband. Among hunter-gatherers, monogamy is common. One reason is that small children must be carried when the group moves its base camp, usually a distance of several miles. This feat would be impossible for a woman whose two children are less than five.

The marriage system is not the same as the reproductive system, however. Thus, legal divorce permits American men to have different wives at different points in their lives. A narrow focus on marriage also misses the fact that the mating system can be more polygynous than the marriage system. Extreme examples are seen in the case of certain basketball stars; for example, Wilt Chamberlain, who claims to have had sexual encounters with thousands or even tens of thousands of different women in the course of his life.[3]

Even in societies where polygamy is freely permitted, it is rare because most men cannot afford to maintain more than one family, even if they happen to share a single home. As a consequence, even though some polygamy was permitted in most societies of the world, the vast majority of the world's marriages were and are monogamous.[4] Although monogamy is legally enforced in the United States, the Mormons practiced multiple marriage in Utah until the practice was suppressed by the United States (see fig. 8), though a minority still practices it.

On the basis of marriage practices, humans can therefore be classified as "almost monogamous" or "mostly monogamous." The trouble is that, to a biologist, we don't look monogamous! In monogamous mammals such as foxes and gibbons, males and females look similar and are about the same size. Differences in size are a rather good indication of the breeding system. Males are often larger than females because they fight with each other for sexual access to females. Where this competition is very severe, as in the case of elephant seals, sex differences in size are enormous. Elephant seal bulls may be more than four times the size of cows and are so heavy that cows are occasionally crushed to death in the course of mating.

Men are typically one-tenth taller than women. Given that weight increases in proportion to height raised to the third power, this predicts that men should be approximately 30 percent heavier than women, and that is approximately what we find. This similarity in height-weight relationship between the sexes masks an interesting sex difference in body composition. Men have approximately twice as much musculature on their upper bodies, which can be seen as an adaptation for hunting or for fighting with other men. In all probability, both functions are involved. The reduced musculature of women's bodies allows them to store more fat, which is critical to the success of reproduction and lactation. The substantial but not extreme difference in body size suggests that, as a species, we should be mildly polygynous, which is exactly what the behavioral and psychological evidence indicate.

Another clue to ancestral mating arrangements is provided by the size of men's testicles. Male primates in species with relatively large testicles experience more mating competition; they produce a large volume of sperm to compete with the sperm of other males present in the reproductive tract of the female. Human testicles are intermediate in size, suggesting that ancestral males engaged in some level of sperm competition. If so, ancestral women were not entirely monogamous.[5]

Fig. 8. Seven of the wives of Mormon leader Brigham Young. Polygamy does not have to be fun. (Reproduced from the Collections of the Library of Congress, LC-USZ62-117487)

Modern genetic evidence also indicates that a substantial fraction of British and American children are fathered outside the marriage, suggesting that some women have a furtive sex life just before, or during, their marriages. It is not possible to put a precise number on this, however, because such data are not published in order to protect the confidentiality of research participants in a highly sensitive area of their lives, where even a hint of suspicion could trigger paternity tests having potential adverse consequences for families.[6] Although it takes two to have an extramarital affair, conflicting reproductive interests of men and women have produced adaptive differences in sexual psychology. These sex differences have important implications for our understanding of violence and conflict in sexual relationships.

SEX DIFFERENCES IN SEXUAL JEALOUSY

Men tend to be very upset by the sexual infidelity of their wives. This fits with evolutionary logic because their reproductive interests are seriously threatened. When a wife is unfaithful to her husband, there is a good chance that the lover may father her child. British biologists Robin Baker and Mark Bellis found that women often timed their infidelities to coincide with ovulation and that they were less likely to use contraception with the lover than with their husband. This meant that the lover was more likely to sire any children conceived at the time of the affair.[7]

A cuckolded husband risks not raising a child of his own and also wastes his paternal investment on the offspring of another man. It would be surprising if our male ancestors had been indifferent to this possibility. If they were, they would have left few offspring of their own to carry on their lackadaisical habits. This helps to explain why sexual infidelity is more distressing to husbands and why even the suspicion of infidelity can, unfortunately, provoke deadly aggression by men against their wives. Note that the emotional response of sexual jealousy would protect paternity even in societies where people did not grasp the connection between intercourse and pregnancy.

From a cross-cultural perspective, violence of husbands against wives is the leading cause of violent death.[8] Some level of violence against women can be regarded as normative for virtually all traditional societies—abhorrent as this may be to us—although its frequency and intensity may vary considerably. Thus, English common law endorsed a rule of thumb according to which a man was permitted to beat his wife with a stick, provided the stick was no thicker than his thumb![9]

Even in our own society, where women enjoy near equality of social and economic status, severe violence of husbands against women is surprisingly common. Although difficult to quantify because it is often concealed by the victim, abuse by husbands and lovers is a leading cause of injury for women between the ages of fifteen and forty-four years. It is estimated that one woman in four will be a victim of physical assault by a partner or ex-partner during her lifetime. American women, too, often use physical aggression against their husbands, and at least one researcher has found that women are violent more often than men. Yet husband battering is a far less severe social problem because fewer women have a murderous intent and because of the obvious fact that men are larger, stronger, and better able to defend themselves.[10]

Violence against women the world over is produced by a psychology, derived from evolution, of male possessiveness and sexual jealousy. Male sexual possessiveness is accommodated by the legal codes of many countries. Marriage, in many countries, is considered to make a woman the sexual property of her husband. Under British law, a husband could sue his wife's lover for "criminal conversation" and be awarded damages. The idea was that the lover had infringed a property right of the husband and was therefore liable for damages. Archaic though the law may seem, it has actually been invoked in the twentieth century to curtail the activities of adulterous male lovers.

By the same logic, adultery was considered an unbearable insult to the husband. If he caught his wife in the act of adultery, according to the *in flagrante delicto* principle, he was justified in killing both the spouse and her lover. This principle was the law of the land in the state of Texas until 1960. In contrast, serious aggression by wives against husbands is most often a response to being threatened, or feeling threatened, in an abusive relationship.[11]

THE TRAP OF AN ABUSIVE MARRIAGE

One of the greatest mysteries about severely abusive relationships is why sane and even highly educated women would remain in the home until they sustained life-threatening injuries. Husbands who go on to be dangerously abusive may initially seem indistinguishable from other men in terms of their sexual possessiveness. Men who beat their wives may be very ordinary in other respects. They may even come across as kind and charming. This impression is particularly likely to be made early in the relationship. By the time a pattern of violence emerges, children may have been produced, and a woman may feel trapped by emotional and financial bonds.[12]

In the past, a great deal has been made of the dependent personalities of victims, as though this were a primary cause of the abuse. Yet even strong and independent women can be broken by a pattern of repeated severe abuse. The sheer prevalence of wife-battering suggests that abusive men are not extremely abnormal, but represent the upper end of a continuum of rampant jealousy.

Spouse abuse often begins gradually and progresses through numerous cycles until it spirals into the husband attempting to kill his wife. Each cycle consists of three phases. In the first phase, tension builds as the wife attempts to keep the husband calm by talking him out of his jealous spells. Then there is an explosive phase with actual battery. This is followed by a period of the batterer behaving in a contrite and loving fashion.

If a woman wishes to leave her husband, her resolve is weakened by such emotional roller coasters. On the one hand, she is battered into submission. (Most experts agree that a primary goal of batterers is to control their victims, and evolutionists would interpret this as stemming from an evolutionary psychology of sexual possessiveness.) On the other, she may be swayed by the man's tender, sympathetic side which conveniently follows an episode of abuse.

Spousal abuse can be compared to brainwashing. This is most effectively accomplished by a combination of terror and kindness. It is an interesting phenomenon that people who are held hostage may come to identify with their captors, as in the sensational case of heiress Patty Hearst, who appeared to mimic the criminals who had abducted her.

As a husband becomes progressively more threatening, his wife becomes psychologically weaker. At a certain point, she effectively becomes an accomplice in the crimes committed against herself. She does this both by accepting responsibility for the husband's actions—even coming to believe that she deserves the abuse heaped upon her—and by helping to conceal the scars and bruises of the battering. She often provides ingenious explanations for the injuries or uses makeup or clothing to hide bruises. At this stage, a woman's inappropriate sense of guilt makes it rather unlikely that she will seek help.

Perhaps the single biggest reason that an abused wife does not leave her husband is the fear that this action will finally drive him over the edge, causing him to kill her or her children, or both. This fear is well-founded. Abusive husbands are needy and dependent as well as controlling. They may live in fear of being abandoned for another man. Most battered wives are well aware that simply moving away will not solve their problems. Thus, women living in shelters for battered spouses are typically harassed on a daily basis by their ex-partners—often for more than a year.

Physical abuse is always accompanied by psychological abuse. The battered woman becomes virtually a hostage in her own home. Her psychological enslavement means that abstract principles of legal equality cannot reach or help her. Ironically, women in a preindustrial world were unlikely to be so vulnerable to extreme abuse because they could appeal to close relatives to protect them. The isolated conditions of modern urban life provide a context in which women may be particularly vulnerable to the sexual jealousy of their husbands. On an emotional plane, a battered wife is stuck in a pretechnological world in which men use their physical strength to control and subjugate women. Depressing as it may seem, there is good reason to believe that the sexual jealousy of men has always driven them to behave in this way. This does not excuse spousal battery, but it does make the case that it is an extreme manifestation, in modern environments, of the evolution of emotional responses of men which helped them to strive for exclusive sexual access to their wives. In a mildly polyandrous mating system where wives were occasionally unfaithful to their husbands, this would have increased the husbands' reproductive success and guaranteed that the trait of sexual jealousy was present in future generations of men.

Many social scientists have assumed that the tendency of men to treat women like their sexual property is due to greater masculine economic and political power. Yet in societies in which women control property, there is no evidence of violent sexual possessiveness on the part of women. In about 17 percent of human societies, the husband takes up residence close to the bride's family (*matrilocal societies*). In many of these matriarchal societies, descent is traced and property is inherited via the female line. This does not mean that men are treated like property by all-powerful women. Neither do women use property to control men in the same way that men use property to control women.[13] Male sexual jealousy is not simply a response to the powerlessness of women, but reflects the evolution of their psychology. Just as fear motivates people to avoid dangerous situations, so the unpleasant emotion of sexual jealousy motivates men to avoid threats to their reproductive success.

WHAT UPSETS MEN AND WOMEN ABOUT INFIDELITY

Patterns of sex differences in violence and possessiveness can be observed in societies around the world. Moreover, an evolutionary approach to sexual jealousy can be scientifically tested. Since men and women have different vital interests in marriage, the situations that excite sexual jealousy should differ between the sexes. Because a single act of infidelity can be a real threat to a husband's paternity, men should be more upset by the prospect of their wives committing adultery than women are by the prospect of their husbands' philandering. After all, a single act of sexual intercourse can entirely undermine a large chunk of a man's lifetime reproductive effort. On the other hand, an act of adultery by the husband might have no negative consequences for the reproductive success of the wife, although it could be seen as a betrayal of trust and would expose the wife to increased risk of venereal disease.

If the husband's adulterous behavior continued, resulting in the formation of a bond of emotional intimacy, however, the wife should probably be even more concerned. Romantic love is the emotional dimension of a willingness to provide economic support to the marriage. At a certain point, the philandering husband might become so enamored of his lover that he might abandon his wife, choosing to set up house with her and invest all of his material resources in the children of the second household, to the detriment of the children of the first household and their mother.

Consistent with these ideas, David Buss, who is now at the University of Texas in Austin, and his colleagues, found that men are more likely than women to say they would be upset by a spouse's sexual infidelity. Women are more likely to say that they would be upset by indications that their spouse is forming an emotional attachment to another woman, however. While interesting, these results can be questioned because the data are self-reported. It may be that people's intuitions about how much a hypothetical event would upset them differ from how they would respond in a real situation.[14]

Such methodology is equivalent to asking research participants, in some ways, to be their own social psychologists and to draw conclusions about their own likely responses to a particular scenario. This may *sometimes* produce the correct answer. After all, people often do have good insight into their own motives. Yet, there are many reasons why people might have insufficient knowledge of their own motives or might distort their responses to give what they believe is the "right" answer.

Buss and others also measured physiological responses to the different scenarios using the same type of apparatus as used in a polygraph ("lie-detector") test, which measures emotional arousal (rather than lying). Merely thinking about their spouse having sex with a rival produced a much larger physiological response in men than in women. For example, the heart rate increased by an amount equivalent to the effect of drinking two cups of coffee. Assuming that it is much harder to fake this sort of physiological reaction than it is to fake a verbal response, the physiological data can serve, to some degree, as objective confirmation of the self-reports.

Men become extremely upset by the mere thought of their wives being sexually unfaithful, whereas women seem less upset at the thought of their husbands having sex with other women. This seems to indicate that male sexual psychology is shaped by an evolutionary selection pressure which caused them to be sensitive to any threat to their paternity. It also helps to explain why jealous men may be motivated to behave aggressively when confronted by evidence of their spouse's sexual infidelity. In the case of spousal abuse, it illuminates the otherwise inexplicable tendency of jealous men to beat their wives and accuse them of infidelity, even when there is no indication that this has happened. Of course, this does not excuse or justify such brutal and irrational behavior in men; it simply explains where it may come from.

In societies where wives do frequently have extramarital relationships, such as many of the island societies of the South Pacific, men may withhold some or all of their investment in children of the marriage; they thereby protect themselves against wasting their reproductive investment by nurturing the children of another man. In such societies, hostility between husbands and their wives is common.[15]

Despite the research noted earlier, it would be a mistake to assume that women may not be deeply upset by the sexual infidelity of their husbands. They are more likely, nevertheless, to be sickened by a betrayal of trust rather than the fact of sexual intercourse itself.

It is thus misleading to assume that women are completely indifferent to the sexual infidelities of their husbands. Women do feel sexual jealousy, although they are less likely to respond to jealousy with deadly violence. Unfortunately, the evolution of male sexual possessiveness leaves women vulnerable to harassment, coercion, and even rape.

SEXUAL HARASSMENT

Sexual harassment arises from a pervasive conflict between the sexes over desired levels of physical intimacy and sexual access. Conflict between the sexes over opportunities to mate is by no means peculiar to humans. It is a truism of biology (known as Bateman's principle) that males are everywhere more eager to mate and less discriminating in their choice of a partner. In fact, Bateman generalized this rule from his study of the mating of fruit flies. Males increase their reproductive success by inseminating as many partners as possible. For females, having sex with many partners seldom increases their biological success. On the contrary, a female that avoids mating indiscriminately and holds out for the mate who can contribute the best genotype to her offspring favors their survival and her own reproductive success. Needless to say, animals do not understand genetics, but their behavior is shaped by natural selection to promote adaptive mate choice.[16]

In addition to genetic quality, females are sensitive to the ability of a potential mate to invest in offspring. Take the case of scorpion flies, so called because the male genitalia, which are large and bulbous, curve upward in the shape of a scorpion's sting. Female scorpion flies can be induced to mate by the gift of an insect delivered by the male, for example. While the female is busily devouring the nuptial gift, the male is busy copulating. If he is stingy and brings a small fly as a nuptial gift, the female quickly finishes it and copulation is terminated![17]

Viewed as a market transaction, it can be seen that the scorpion flies are trading off female access to food for male access to copulation. The male provides an economic good (the food) in return for sexual access to the female. In principle, this transaction is an exact parallel with the prostitution trade among humans. Prostitutes are almost always women because sexual access is a limited resource that women—not men—control.

This is the conclusion drawn by Donald Symons in his review

of human sexuality across the cultures of the world. He notes that it doesn't matter how unattractive a woman is believed to be, she never has any difficulty in obtaining sex and becoming pregnant. He thus argues that for humans, as for other species, sexual access is a female favor that is granted to men. The desire for sexual intimacy is higher up on men's agenda for romantic relationships, whereas the desire for emotional commitment is higher on women's agendas.[18]

These different agendas may lead men and women to interpret the same social situation in different ways. In one experiment, participants were asked to interpret a video clip of a woman greeting a man. Women interpreted this scene as being largely devoid of sexual intent. The woman was just being friendly. Men were much more likely to interpret the scene as having a sexual connotation. For them, the woman greeted the man because she was interested in having a sexual relationship with him.

This fits in with the evolution of differing male and female approaches to sex. Whereas men are favorably predisposed to casual sex, women in general want to avoid it. If a man assumes that a woman is sexually interested in him, and he is correct, then he can capitalize on the opportunity. If he draws the (more realistic?) conclusion that she is merely being friendly, an opportunity for casual sex may have been lost. Adaptationist thinking predicts that under conditions of uncertainty, men may be predisposed to infer sexual interest.

The female desire to avoid casual sex explains why women interpret the videotaped social interaction very differently from men. Sociability is an important quality for women, one that allows them to meet and attract potential husbands. Greeting a man may mean that a woman is interested in getting to know him on a very superficial level. From there, she can decide if she is interested in a deeper relationship. It certainly does not mean that she is even remotely considering having sex with him. For this reason, it is adaptive for women to make clear distinctions in their own minds between being friendly and being sexually provocative.

Such conflicts of agendas and perceptions have arisen in the workplace and have taken on increased significance with the entry of women into previously male-dominated occupations. The sexual assertiveness of men can make the work environment difficult for women, which has necessitated the enactment of sexual harassment legislation. Before women were protected by laws, men responded to the few attractive female employees in their midst by treating them as sex objects, telling obscene jokes, making indecent proposals, and taking hands-on liberties. These actions are generally very upsetting to women because they are fundamentally coercive and tend to undermine a woman's desire to be highly selective in the choice of a sex partner. They are also insulting, in the sense that they tend to undermine the woman's sexual reputation by implying that she is common sexual property. Furthermore, in the hostile environment created by sexual harrassment women are less likely to be taken seriously as professionals, thereby damaging their ability to carry on their work and hence their prospects for pay raises and promotion. If such a situation persists, they may also feel obliged to quit their jobs and seek more humane work environments with adverse consequences for salary and promotion.

With the advent of sexual harrassment laws and the use of training designed to protect companies against lawsuits, women today are treated with greater sensitivity. We could say that there has been a reversal, in which the male perception of reality has been legally supplanted by the female one. If this approach had not been adopted, it would have been impossible to get rid of even obvious sexual harrassment in the workplace.

If the two sexes are proceeding under radically different agendas and interpreting the same events in different ways, then it is not surprising that misunderstandings can occur in and out of the workplace. It is theoretically possible for misunderstandings to occur about the desired level of intimacy between a man and a woman on a date. With this in mind, Vassar College adopted a code regulating dating behavior of students, according to which no intimacy may occur unless it has first been explicitly agreed to in a sort of verbal contract. While this has amused many observers, it does at least address one potential cause of date rape, namely, poor communication between dating couples. It appears that in the ambiguity of certain situations, males tend to infer levels of consent that are not explicitly given, which could cause a problem. On the other hand, a woman may believe that she has drawn a clear line in the sand when she has not explicitly done so. The Vassar code was designed to minimize the problem of consent being falsely assumed and of intimacy proceeding beyond the comfort zone of one partner, usually the woman. This is not to say that date rape is often, or usually, due to a misunderstanding. Too often men have overpowered or drugged their date in order to rape them.

Sexual harassment laws and guidelines might seem designed to stamp out romance in the workplace. That is probably not, however, what women desire. Some romantic overtures on the job are much more acceptable than others. Students of sexual harrassment have discovered that a woman does not consider that she has been sexually harrassed if she receives a romantic overture from a man whom she would consider an appropriate marriage partner. Thus, if she is asked out on a date by an executive of her own rank or higher, who is not her boss or superior, she is very unlikely to complain even if she is asked repeatedly and repeatedly declines. On the other hand, the same overture made repeatedly by a janitor is likely to be seen as harassment, even though the janitor has no power in the company that could threaten the woman's promotion prospects. Evolutionary psychology is useful in predicting which situations a woman will most likely interpret as sexual harrassment. They are situations which tend to undermine her evolved selectiveness in the choice of a mate.

Sexual harrassment is thus defined in very subjective terms. A woman is harrassed only if she feels that she has been harrassed. From a legal point of view, asking a female coworker out on a date is not sexual harrassment. It becomes harrassment after repeated unwelcome overtures are made, particularly if these are accompanied by any type of coercion, or if these unwanted overtures make the work environment unpleasant. One fascinating aspect of the legal definition of harrassment is that women are not at all put out by masculine persistence—provided they perceive the coworker as an attractive, socially acceptable, dating partner. Many women are flattered when men they find attractive repeatedly ask them out, and such persistence is sometimes rewarded. Women may see repeated requests for a date as evidence of their own desirability and of constancy and commitment on the part of the suitor. Men who are interested only in sexual gratification do not have to be refused very often before they move on to greener pastures. From an evolutionary perspective, women are liable to charge men with sexual harrassment when an undesired coworker uses the work setting to make sexual overtures that conflict with her evolved mate-selection criteria.[19]

MALE SEXUAL PSYCHOLOGY AND RAPE

Given the enormous legal penalties exacted on rapists, it is a surprisingly common crime. It is estimated that 15 percent of college women experience date rape. Fourteen percent of wives are raped *by their own husbands*. About one woman in four has been sexually abused as a child. The lifetime incidence of rape may be as high as one woman in three. The tendency to commit rape is apparently more common than we would like to believe. When asked whether they would commit rape in a setting where there was no possibility of adverse consequences, approximately one-third of American college males admit that they would commit this crime. Since it is very socially undesirable to admit to such a propensity, it can be suspected that the true number may be considerably higher.[20]

What if the evil propensity to commit rape were due to genes? Men who commit rape would be inseminating women. If this contributes to their reproductive success, then the number of copies of "rapist" genes would increase in future generations. The bottom line would be that all men are rapists because their genes were making them do it. This is a chilling thought; however, it is not scientifically valid for a number of reasons.

Forcible sexual intercourse is a rather common reproductive strategy among males of other species, but not among humans. In the case of ducks, males and females pair up during the breeding season in what appears to be the monogamous relationship seen in 60 percent of birds. However, the males are merely guarding females from the sexual encroachment of neighboring males who are always willing to engage in forcible copulation. We know this because as soon as the eggs are fertilized, the male deserts and attempts to set up another family.

Forcible intercourse is taken a step further by scorpion flies. The male scorpion fly may use different mating tactics. The consensual approach involves "buying" sexual access with a piece of food, the nuptial gift. Alternatively, the male may bind to the female with specialized claspers and inseminate her without her consent. These claspers have no known function outside of copulation and when they are experimentally inactivated, forcible intercourse is no longer possible. In the case of scorpion flies, rape is a reproductive strategy and this explains the specialized clasping appendages.[21]

Forcible rape occurs also among primates. It is fairly common for orangutans. The rapist is usually a subadult male and these attacks can be quite violent, with the female being struck and bitten. Females normally consent to mate only with fully adult males. There is controversy about whether forcible copulations can produce offspring, however.[22]

Men do not have any specialized adaptations of anatomy or physiology designed by natural selection to facilitate rape. Neither is there any evidence that sexual predation could be a viable reproductive strategy in men as it is, for example, among ducks, although this case has been made in a recent controversial book by Randy Thornhill.[23] A better understanding can be derived from looking at the potential reproductive benefits compared to the costs of rape. In the normal course of events, the costs are too high and the reproductive benefits too low for rape to be a viable strategy. This calculus is reversed during war and social upheaval, however.

Instead of seeing rape as a reproductive strategy, it is probably better to view it as an extreme manifestation of the sexual assertiveness that is typical of men. In all societies, rape is treated as a serious crime. In our own, it is roughly equal in seriousness to manslaughter, carrying a potential life sentence. Many subsistence cultures have taken a sterner approach: the rapist is killed by irate relatives of the victim. Since a fertile woman may have unprotected sex scores of times without becoming pregnant, the likelihood of pregnancy from a single act of rape is very low. Given that the reproductive costs were normally much greater than the potential reproductive gain—because rapists were likely to die young and childless—rape could not be favored as a primary reproductive strategy.

Much of what we know about rape fits in with the view that it can be considered a facultative behavior. In other words, under certain circumstances, the normal pattern of male sexual pushiness may well cross the line into forcible rape. This is a controversial statement, but it is supported by the scientific evidence.

First, it is important to establish that rape is a sexual crime. It is not simply the manifestation of a desire to hurt, control, frighten, or humiliate women, although these motives may certainly coexist with sexual motives. It is true that there is a small minority of sadistic rapes, which receive a lot of publicity, where nonsexual motives apparently predominate. These crimes have no more in common with the majority of rapes than serial killings have with the majority of murders. The perpetrators are a highly abnormal subgroup of the male population.

The view that rape is essentially an act of political terrorism through which all men frighten and control women was advocated by feminist Susan Brownmiller in her book *Against Our Will*.[24] The primary motive for rape is not political but sexual, however. The strongest evidence for this comes from an analysis of the demographics of rape victims. If Brownmiller's view were correct, then we should expect all women, regardless of age, to be potential victims. The data show a very different picture. Young women in their early twenties are much more likely to be rape victims.[25] Data for twenty-six U.S. cities found that women were six times as likely to be victims between the ages of sixteen and twenty-four years as they were at the ages of twelve to fifteen years or thirty-five to forty-nine years. Since the age of highest victimization is also the age of maximal sexual attractiveness, because women of this age are right at the beginning of their plateau of high fertility, it is clear that women are being targeted because of their sexual attractiveness. It could be argued that young women of this age go out more and are more likely to be out late at night, for example, making them more vulnerable to attack and thus providing an alternative explanation for the observed age pattern, but this argument is not supported by the evidence. Other violent crimes against women, such as homicides, did not reveal a similar age pattern. The vulnerability of young women to rape cannot be explained in terms of greater likelihood of encountering the perpetrators. They are selectively targeted because of their physical attractiveness. Although women are raped at all ages, their likelihood of victimization is much lower after the age of forty. This pattern has been observed repeatedly in different studies.[26]

While rape victims are most likely to be young attractive women, who are thus desirable to most men, perpetrators are often highly unappealing in terms of appearance and social status, although men of all social strata commit rape. They tend to be young (although the age range is wide) and to have low social status, which means that highly desirable women are not available to them by lawful means. Since attractive women elicit their sexual desire without the possibility of legitimate sexual access, rapists may feel frustrated and angry and may vent these emotions on their victims.[27]

The role of low social status and consequent rejection by attractive women was stressed by one (repeat) rapist in accounting for his actions:

I wanted this particular type of person—she was a college girl—but I felt that my social station would make her reject me. And I just didn't feel that I would be able to make this person. I didn't know how to go about meeting her. Anyway, I waited one night until she had gone to bed. After the lights were out, I just went into the window. She was frightened, of course, and I took advantage of her fright and raped her.[28]

This account emphasizes the desirability of the victim and her inaccessibility to the rapist by socially acceptable channels. It suggests that rape is more likely among men who have no other options for having sex with highly desirable women. Such men would be least sensitive to the very great costs associated with the crime and most sensitive to its benefits. For men who normally have more sexual access to women, rape is likely to occur only when the costs are very much diminished, something which often occurs with the collapse of civil order during warfare. The motivation for rape within marriage is obscure, however. It might reflect a conflict of interest over frequency of intercourse or it might indeed be perpetrated to humiliate and terrorize the victim.

Evolutionary psychology is not only useful in explaining the characteristics of rapists and victims and the circumstances under which rape is most likely to occur (thereby providing indispensable information for those who work to prevent rape), but it also helps us to understand the feelings of the victims. A single woman who is raped may experience a severe loss of desirability as a bride in many cultures of "honor." Even if she does not become pregnant, her social standing and ability to make a desirable marriage can be severely compromised. In that sense, something is taken away by force that cannot be replaced.

This also applies to the psychological impact of rape. Rape victims are likely to experience posttraumatic stress disorder. It is estimated that some of the symptoms of posttraumatic stress show up in 95 percent of women who have been raped within two weeks of the attack. The symptoms include difficulty in sleeping, flashbacks to the traumatic event which occur during sleep in the form of nightmares, constant vigilance and anxiety, difficulty resuming normal sexual behavior, and a general sense of numbness and detachment from other people that can disrupt friendships and other intimate relationships. The severity and persistence of the symptoms is related to the violence of the attack. After six years, three-quarters of the victims say that they have recovered. However, for about one-sixth of victims, posttraumatic stress syndrome is still present seventeen years later and has evidently developed into the chronic form seen in combat veterans. The psychological effects of rape can therefore be as great as any other stressor to which a person may be exposed in their lifetime.[29]

PORNOGRAPHY

Male sexual psychology is designed for the mild polygyny that is characteristic of our species, as indicated by body-size differences between the sexes and other adaptations, such as testicle size relative to other primate species. Mating competition is also promoted in men by mechanisms that are largely psychological. For example, when married couples are separated for several days, the man produces an unusually high volume of sperm in his ejaculate following reunion. Absence makes the heart grow fonder, and the greater production of semen reflects the more intense sexual excitement experienced by the husband following a separation. This excitement probably has a very pragmatic basis in evolutionary terms. When a man was separated from his wife or lover for an extended period, there was a reasonable chance that she could have had sex with another partner. The increased volume of sperm is designed to compete with that still present in the woman's reproductive tract following copulation with another partner or partners.[30]

On a psychological level, the man is very glad to see his wife and that promotes sperm competition. Of course, men are sexually excited by other women apart from their wives and the same mechanism comes into play. When a man meets an attractive woman, he is sexually excited, which produces an increased volume of sperm even though there is no sexual interaction. If reproduction is thought of as a war fought between men, then he is busy stockpiling weapons.

It has often been pointed out that it takes two to tango and that men cannot be more polygynous on average than women are. Interestingly, when college students are asked about the number of their sexual partners, men claim to have had twice as many partners as women. These numbers could work out only if a small number of highly sexually active women, such as prostitutes, took up the slack for the rest of their sex. One researcher has concluded that it is impossible for prostitutes to be this busy! The data are thus unreliable. In all probability, both sexes are lying. Men exaggerate the number of their sexual conquests and women understate theirs. Of course, there may also be some ambiguity as to what constitutes a sexual encounter. Former president Bill Clinton is not the only person who is confused about whether he had sex with that woman! The diametrically opposite reporting biases of men and women reflect another basic psychological difference. Even in the supposed sexual egalitarianism that reigns in our society, men see having many different partners as generally desirable whereas women do not.

One manifestation of the masculine interest in sexual variety is the pornography industry. Pornographic enterprises generate over eight billion dollars annually in the United States alone, and almost all of this money comes from men.[31] Attempts to interest women in depictions of nude men by the magazines *Playgirl* and *Viva*, founded in the 1970s, proved to be an instructive and amusing failure (recounted by anthropologist Don Symons in his 1979 book, *The Evolution of Human Sexuality*).[32] Both enterprises seemed to flourish, but reader surveys revealed an embarrassing fact: the majority of readers were gay men!

Symons makes much of the fact that gay men are so visually oriented, pointing out that, apart from the object of their sexual attraction, homosexuals have a masculine sexual psychology in all other respects. Thus, there is no interest in publications such as *Playboy* on the part of lesbians. Like heterosexual women, lesbians have trouble understanding how men could be sexually excited by the exaggerated and implausible sexuality depicted in pornography. It seems that there is no market for depictions of nude women designed to stimulate lesbian fantasies or, at least, no such commercial magazine has been produced.

The agreement of heterosexuals and lesbians in their reaction to pornography suggests that women are simply not excited by viewing erotica. Laboratory studies that measured physiological arousal using a plethysmograph, which measures blood flow to the genitals, found, however, that maximal excitement could be produced in women by viewing pornographic films.

Although women are thus highly responsive to erotica, they are not motivated to seek out this experience. According to Symons, this reflects a basic sex difference in evolved sexual psychology. While men's sexual desires are fairly closely approximated by pornography, with its focus on sheer physical gratification in the company of any of a number of enthusiastic and physically attractive strangers, women's desires are more closely approximated by the romance novel. Here, sexual expression is depicted as the fulfillment of a deep emotional attachment, a love relationship.

Men's craving for impersonal sex, a craving which may never be expressed, reflects an evolutionary past in which casual sex could have greatly increased reproductive success. In contrast, women would not have had more children by having more sex partners. Moreover, they would have been more successful in raising healthy children if they focused their affections mainly on one man, since this would give the man confidence that the children of the marriage were his and he would be more likely to support them. Pornography depicts sex between strangers, whereas women are more interested in sex with men they love.

SEXUAL FETISHES

Women sometimes express astonishment that men could become sexually aroused by viewing pictures of women they will never meet or have any interaction with. In some cases, the button of sexual arousal is pushed by the depiction of a body part rather than an entire woman. Thus, some pornographic magazines specialize in the depiction of women with gigantic breasts. Some men have an intense fascination for feet, others for hair, legs, or buttocks. When this passion takes over their erotic lives, it is referred to as a *fetish*. Even though fetishes may become so strong as to be distinctly abnormal, research has shown that normal men are liable to develop a sexual interest in any object that is reliably paired with sexual excitement. We know less about the extent to which normal women may develop fetishes. Consistent with their lack of interest in sexual variety for its own sake, women are reluctant to volunteer for research dealing with sexual arousal, but they might also feel more dubious about the motives and ethics of sex researchers. The clinical psychology literature finds that women are far less likely to report fetishes. These typically develop later in life and appear to be less stable than those of men.[33]

One scientific answer to the question of how fetishes develop is to create a fetish in the laboratory. The project of turning normal young men into sexual deviants is clearly an ethical problem which the researchers addressed in their own way.

The inspiration for the 1966 experiment of S. Jack Rachman and Ray J. Hodgson, of the Institute of Psychiatry in London, was Pavlov's research with salivating dogs. Pavlov's idea was to make the dogs salivate to something completely unrelated to food (a buzzer) by presenting the buzzer just before the food.

Instead of measuring drops of saliva, the intrepid researchers measured sexual arousal in men directly using a penile strain gauge. This is an ingenious device in which the stretching of a rubber tube placed around the penis alters the electrical conductivity of the column of mercury it contains allowing for continuous automatic measurement of sexual arousal. First, the researchers presented a slide of the intended fetish object, a pair of black fur-lined boots (chosen because fetishes about articles of clothing are common). Just to be sure that none of the participants in the experiment were already erotically interested in footwear, they were exposed to slides of the fur-lined boots. None responded. All were normal.

Now the researchers proceeded with their sinister plan. A slide of the black fur-lined boots was repeatedly followed by a slide of an attractive nude from a pornographic magazine. After twenty such pairings, most of the men were sexually aroused by the slide of the boot. The experiment was a complete success—all of the normal participants had been turned into fetishists.

One might imagine that the researchers would be a little stunned by their own success, but they were also mindful of the ethical dilemma. Using Pavlov's work as an inspiration, they devised a solution. They presented the boots slide repeatedly, without the nude—analogous to Pavlov repeatedly sounding the buzzer without food—and found that the fetish slowly went away.[34]

This would have been a happy story were it not for the fact that the researchers decided to contact the same participants again a year later to pursue their studies. When they showed them the same black fur-lined boots, the response was unanimous: strong sexual arousal. Pavlov noted the same phenomenon in his salivating dogs and referred to it as *spontaneous recovery*. The researchers papered over the cracks by having the participants undergo the same procedure for getting rid of the fetish a second time. Had they really created monsters? We will never know.[35]

Most people do not acquire their fetishes from psychology experiments. The research suggests that any object which is repeatedly paired with sexual arousal can acquire the ability to produce sexual excitement through Pavlovian conditioning. This helped our male ancestors to become swiftly aroused in circumstances where sexual intercourse was likely to occur. In addition to the inbuilt tendency to be aroused by fertile and healthy members of the opposite sex, men thus learn to be excited by arbitrary events predicting that they are going to have sex.

A few concepts in evolutionary psychology are thus very helpful at helping us to understand the differing sexual psychology of men and women. Evolutionary ideas are also very useful in explaining why some marriages are stable and others fall apart.

Chapter Six

Why Marriages Fail

Around the globe, marriage is a cooperative arrangement between a man and a woman for the purpose of raising children. It is based on an unwritten contract. Men agree to use their economic resources to support the family. In return, they are guaranteed paternity of the children through exclusive rights of sexual access to their wives. Dated though this view of marriage may seem in an age in which women are reaching economic equality with men, it provides a good understanding of when and why marriages break up: the evolved psychology that men and women bring to their marriages has not changed.

Children as Cement in a Marriage

If the purpose of marriage has primarily been to raise children, then childless marriages should tend to fall apart. At least that is the expectation derived from study of the reproductive behavior of birds. Most species of birds are monogamous during the breeding season and some couples pair up year after year. If they fail to reproduce in a particular season, however, they are likely to "divorce" and try their luck with another mate.

People are not birds, and you would imagine that we are much better equipped to make informed decisions. For one thing, raising children takes a lot of effort and is very expensive, not to mention its emotional price. For parents, every single day is replete with harrowing episodes including tantrums, mysterious illnesses, domestic mayhem, visits to the emergency rooms of hospitals, and disturbing calls from teachers, not to mention the endless hassles associated with meals, tidying, and scheduling.

Despite these responsibilities and demands—and in the face of all rational prediction, but consistent with the behavior of birds—couples with children are more likely to stay together. According to data collected by the United Nations for many countries around the world, children are a powerful cohesive force in a marriage. Couples with no children are twice as likely to divorce (39 percent versus 19 percent) as those having two offspring. Childless couples are more than ten times as likely to divorce as those having four or more children.[1]

Children bind couples together and childlessness drives them apart. Whether they are miserably unhappy and stay together for the sake of the children, or remain married simply because they cannot afford to live separately is another story. In either case, it is clear that the shared responsibilities of raising children keeps couples together, just as success at reproduction cements the pair bond for monogamous birds. Conversely, for couples who do not have children, the natural cooling of their relationship over time may mean that they have no compelling motive for staying married.

THE SEVEN-YEAR ITCH

When birds pair up, they are usually committed to each other for the duration of the mating season. Yet if they fail to produce live chicks, they "remember" this outcome and refuse to join their fortunes with the same mate on the following season. Even though marriages can break up at any stage, there is a pattern, referred to as *the seven-year itch*, according to which marriages tend to become unstable after that time.

This phenomenon, popularized in the Marilyn Monroe movie *The Seven Year Itch*, is based on the notion that the fabric of marriage gets frayed threadbare after seven years. Do a lot of couples actually part company after their septennial?

According to data published by the U.S. Department of Health and Human Services, first marriages ending in divorce actually last an average of eleven years for both men and women, but this number is inflated by a small number of lengthy marriages. After seven years of marriage, 50 percent of divorces have occurred, which supports the popular conception of the seven-year itch. Remarriages ending in divorce lasted an average of 7.4 years for men and 7.1 years for women, possibly reflecting the natural course of emotional cooling in a romantic relationship. The same seven-year itch also shows up in other nations of the world according to United Nations data.[2]

Evolutionists feel that the seven-year span is to humans what a breeding season is to birds. Apparently, natural selection has designed us to withstand the rigors of marriage for at least long enough to raise a child to the point that it has a reasonable chance of survival without a father. After seven years of marriage, the first child is likely to be five or six years old. By this age, children are strong enough to walk without being carried, which is in keeping with constraints on our hunter-gatherer ancestors who had to relocate their temporary camps by distances of several miles. Children at five or six years can feed themselves to some extent, by collecting plant food, for example, and are alert to dangers of wild animals. That said, there is nothing magical about the seven-year period. Couples can and do part at any time. Thus, marriages often break up in the first year due to various issues of sexual, emotional, and domestic incompatibility. This pattern is particularly common for Moslem societies in which there is a wedding dowry that is 50 percent returnable if the marriage is not consummated. Moreover, in-laws may decide to return a new bride if she is not fitting in well, and this is likely to happen early in the marriage.

The fact that marriages are more vulnerable to dissolution at some times than others does not imply any inevitability to divorce. In some societies, there is virtually no divorce. The whole argument revolves around the timing of divorce for marriages that are intrinsically unstable for one reason or another. Most marriages that are exposed to these difficult patches are strong enough to withstand them. It is also true that the quality of marriage changes over time, becoming less dependent on physical intimacy and more dependent on companionship. Marriages that are incapable of making such transitions may be more vulnerable to falling apart.

THE MARRIAGE MARKET

The importance, quality, and permanence of marriage vary greatly across different societies and even within a particular society, these may change markedly over time. For example, the annual divorce rate in England in 1911 was one per thousand compared to fifteen per thousand in 1970.[3] A similar trend has been seen in other industrialized countries. Such extreme changes are often attributed to arbitrary "cultural" trends but they can be accounted for more objectively in terms of the marriage market.

When there is a scarcity of marriageable women, their romantic preferences dictate sexual practices (although in a complex and roundabout way) and respectable women rarely indulge in extramarital sex. When there is a scarcity of marriageable men, male romantic preferences are expressed by their having casual sexual relationships with different women. It is odd that when women have the upper hand in the marriage market, sex roles are traditional (because women have to seem chaste and obedient to increase their desirability as marriage partners), and they thus have little social, economic, or political power. Conversely, when the economic and political power of women rises, they have little control over the marriage market and may find themselves cohabiting or in unsatisfactory, unstable marriages.

Marriage is a market to which men and women respond in predictable ways. These responses reflect sex differences in adaptations for increasing reproductive success. Women's agendas are much more focused on providing a good environment in which to raise children, whereas men's agendas are more focused on maximizing the number of children produced; they strive to have sex often and with different women. Due to this sex difference in evolved motivations, the sex that has more control over the marriage market drives sexual attitudes and behavior, and this affects the relative stability of marriage. When there is an excess of marriageable men, and women therefore control the marriage market, chastity is seen as highly desirable in women because women who lose their sexual reputations have trouble marrying. Marriage tends to be highly stable, and masculine investment in children of a marriage is both high and reliable. The key to the apparent paradox of controlling the marriage market, but having little social power, is to be found in competition among women over the most desirable husbands, those who are able and willing to support them and their children. To win such men as husbands, it is necessary to seem both sexually reserved and obedient because these qualities provide a potential husband with some assurance of paternity of the children.

Conversely, when there is an excess of marriageable women over marriageable men, sexual behavior becomes liberated, premarital intercourse is common, and marriage becomes unstable.

These principles explain why there is so much variability in sexual behavior, in relations between the sexes, and in divorce rates both over time and in different societies.[4]

Across the world's societies, there is great variation in sexual mores. Most communities place restrictions on sexual expression, particularly for unmarried women. These restrictions can be extreme, as in the case of the Chinese, where a history of premarital sex is enough to foreclose marriage opportunities for men as well as women. On the other hand, among the Muria of India, adolescents live in a group home, or ghotul, where the expression of all manner of sexual impulses is actually encouraged before moving on to the sober responsibilities of marriage.[5]

Variation in sexual mores is related to the marriage system. In societies where young women are competing for highly desirable (wealthy) spouses, this can lead to extreme practices, such as claustration, where virgins are perpetually locked up and guarded by kin as is true of high-caste Indians. Such practices are more likely among wealthy families that are preoccupied with obtaining high-status husbands for their daughters. In Africa, infibulation is still widely practiced. In this procedure, the genital opening of women is sewn up, which prevents them from having premarital intercourse. This barbaric operation is a testament to the importance placed on premarital chastity in these societies. Although frequently considered torture, infibulation is practiced by female relatives of the young woman who view it is necessary because it increases the confidence of paternity of a potential husband, and thereby increases the marriage market value of the woman and the prestige of her family.[6]

Where women rely heavily on the wealth provided by husbands to raise their children, premarital chastity is expected. If emphasis on premarital chastity is extreme, as in China, it may extend to men as well as women, presumably because it takes two to tango and male licentiousness may undermine female chastity.

Where women do not rely heavily on the economic support of

husbands to raise children, their sexual expression is relatively free. This is true of most hunter-gatherer societies where the lion's share of subsistence is provided by women. Marriages are correspondingly unstable. Among the Ache of Paraguay, for example, the average number of "marriages" for a person is around eleven, although some anthropologists would like to make a distinction between informal sexual relationships and marriages that are arranged between the families.[7]

THE CHANGING DIVORCE RATE

Social scientists often invoke morals, values, and norms to explain why divorce rates are so high in one society, or at one time in the history of a particular group. Such notions are fairly primitive as scientific explanations because they do not really explain such critical issues as why Americans in 1976, for example, were ten times as likely to divorce as Americans in 1896, to cite a period of exceptionally rapid change in the divorce rate. Saying that divorce was more socially acceptable in 1976 than it had been in 1896 is undoubtedly true, but assuming that the divorce rate increases because of altered values and norms is very shaky inference.

The critical weakness in "values" explanations is that they do not help us to predict how people will behave. For example, there has been a huge increase in the number of African American women giving birth outside wedlock in recent decades. Yet when they were surveyed concerning their attitudes to marriage, these women claimed that they saw marriage as being just as desirable as other American women did. Invoking values does not help us to explain the unusually high rates of nonmarital births among African Americans.

A similar argument can be applied to rising divorce rates. Few of the people getting married today want their marriages to fall apart and when they do, many see divorce as undesirable in itself but the only way out of an unhappy marriage. Rising divorce rates are also attributable to a weak marriage market for women, which makes men less committed to their marriages. One major influence is the abundance of single women who are sexually active before marriage.

Most of the children being born in America today will live in a household from which one of their natural parents has been removed by divorce. This gloomy prediction is based on trends in the annual divorce rate which increased from eight per one thousand married women in 1920 to twenty per one thousand in 1975, and has stayed at this elevated level for the past quarter century.[8] Why are Americans today apparently less content with their marriages than they were at the beginning of the century?

According to sociologist Marcia Guttentag and social psychologist Paul Secord, authors of *Too Many Women: The Sex Ratio Question*,[9] high divorce rates today are due to an oversupply of marriageable women relative to marriageable men. This is a far cry from the days of the frontier when there were far more men than women. Women were so scarce in frontier regions that men who succeeded in acquiring wives counted themselves lucky and were motivated to do everything to ensure the stability of their marriages.

In societies with too many women, by contrast, marriage loses its value for men, making it more unstable and hence less valuable to women also. Since there are more women than men, it is easy for men to get married. Moreover, if a first marriage fails, they can easily remarry. Even more destabilizing to marriage is the pool of young unmarried women who are willing to have sexual relationships with men, which makes the bachelor life much more attractive than it is in societies in which there is little premarital sex.

In low-sex-ratio societies, evolved sex differences in the psychology of romantic relationships come to the fore. Men have more opportunities to satisfy their desire for a variety of sexual partners, which conflicts with the needs of women for emotional commitment. There is a good deal of hostility between the sexes over their sexual relationships. The flavor of this conflict is well captured by Geoffrey Chaucer in *The Canterbury Tales*. Chaucer was writing during a period when there was a scarcity of marriageable men, and consequent marital instability, in England in the fourteenth century. Hostile relations between the sexes are depicted in the combative figure of the Wife of Bath, who had sent five husbands to early graves. The conflict between the sexes obviously extended to the bedroom, and Chaucer followed an antifeminist tradition in assuming that women were sexually faithless. Thus, in "The Miller's Tale," an ingenious young apprentice carpenter convinces his boss that the world is about to come to an end purely to contrive an opportunity to be alone with the carpenter's lecherous young wife. The assumption is made that if this opportunity is presented, the wife will inevitably cheat on her aging spouse.

Since it was relatively difficult for women to marry in Chaucer's time because men were scarce, more women were sexually active outside marriage. Unstable marital relationships meant that women were more likely to be abandoned by their husbands and to be stuck with the financial burden of providing for their children. For this reason, they wanted to become involved in businesses and trades, although this was vigorously opposed by the male-dominated trades guilds of the day. The economic difficulty in the lives of women of the period was responsible for the emergence of a feminist movement in literature, as represented by the writings of Catherine de Pizan. All of this is strikingly reminiscent of the emergence of a vociferous feminist movement in America at a time when the supply of men of marriageable age was falling.

THE MARRIAGE MARKET

Sexual intercourse is often seen as a service that women provide to men, but marriage is a different story. Anthropologists see marriage as an arrangement in which husbands and wives combine their economic and social resources for the purpose of raising children. In some societies and at some historical periods, wives are more in demand than husbands. In others, husbands are more in demand than wives. This is often reflected in premarital transfers of property. In some societies, men must pay a price to the bride's family before the marriage is formalized. In others, the woman's family pays a dowry that allows her to marry up the social scale. This system applies in India's caste society, where men from noble families are in great demand as bridegrooms and command huge dowry payments from the bride's parents, who want to establish her in a desirable home. Marriage is thus a true market system that follows rules of supply and demand. This market affects the quality and durability of marriages.[10]

In the marriage market, the sex that is most in demand sets the rules, and this has pervasive consequences for relations between men and women in a society. By analogy, if there is a strong demand for Van Gogh paintings, then the person selling the painting has more control over the price than the purchaser does. Even though all markets are based on eventual agreement, the dynamic of a market usually favors the interests of one individual over another. Thus, when brides are in greater demand than bridegrooms, women get the kind of marriages that satisfy their evolved romantic needs. With a scarcity of marriageable men, the marriage system satisfies masculine romantic needs and often fosters some type of polygyny.

In warlike subsistence communities, like the Mundurucu of Brazil, men are likely to fall in battle and the resulting scarcity of adult men means that if all women are to marry, many must inevitably share a husband. Polygynous marriages satisfy the evolved sexual needs of men. Some polygynously married women are said to enjoy the social companionship and economic cooperation of co-wives. Yet there are undeniable adverse effects for some women. Typically, the senior wife has special privileges and may exploit the labor of junior wives. Moreover, among the Dogon of Mali, children of monogamously married women have much higher survival rates than children of co-wives, suggesting that there is a very dark side to reproductive competition between women in polygynous households.[11]

The other extreme of too many men manifested itself in the American frontier where hard living in the wilderness appealed more to men than it did to women. There was a peculiar double standard in which women were seen either as madonnas or whores. Most men could not marry and had to rely on prostitutes as a source of sexual gratification. Married men considered themselves lucky and were devoted to their wives, so that marriages were highly stable. Despite the difficulty of frontier life, women's evolved romantic need for emotional commitment from their spouse was being fulfilled. The same need is expressed among modern American women, despite an active premarital sex life with several partners. In spite of their sexually liberated behavior many young women feel romantically unsatisfied, in sharp contrast to their boyfriends.[12]

Ideally, the marriage market would be measured as the ratio of single women to single men during the peak ages of marriage, which are two to three years earlier for women than men.[13] Difficulty of marriage for both sexes is generally a function of the overall proportion of males to females in a society. The population sex ratio, or number of males of all ages per one hundred females of all ages, is often a rough proxy of the marriage market. By convention, the ratio is always computed as the number of males over the number of females. If there are more males than females, the sex ratio is greater than one hundred and is referred to as "high." Conversely, when there are more females than males, the ratio is below one hundred is referred to as "low."

THE MARRIAGE MARKET AND DIVORCE

The divorce rate is almost perfectly predictable from the proportion of men to women in the population. In 1920, when the annual divorce rate was eight per one thousand married women, there were 104 males for every female. By 1980 the tables had turned, and there were only ninety-five males for every one hundred females, and the divorce rate had risen to twenty-three per one thousand married women. The sex ratio has remained virtually unchanged for the past thirty years and is paralleled by a steadily high divorce rate. The correlation between the population sex ratio (or number of males per one hundred females) and the divorce rate at four-year intervals between 1896 and 1992 was –.91, indicating that changes in the number of men relative to women accounts for 83 percent of the changes in divorce rate. These numbers are only approximate because they do not look at the ratio of men to women at the peak ages of marriage. An ideal measure of the marriage market would look at the ratio of unmarried men to unmarried women at the peak age of marriage (which is about two to three years earlier for women). Moreover, the numbers are not corrected for the proportion of men and women who are homosexual. Another potentially important influence is the proportion of men in stable, well-paid employment. If all of these influences on the marriage market were taken into consideration, it is possible that changing divorce rates could be predicted with high accuracy from changes in the marriage market.[14]

The sex ratio alters divorce rates in part by changing the quality of the marital relationship. Guttentag and Secord noted that in low-sex-ratio societies, i.e., where there is an excess of women, there is hostility between the sexes combined with a lack of commitment in romantic relationships. Lack of commitment on the part of men stems from their mildly polygynous psychology. Even though they may be devoted to their wives, they are sorely tempted to satisfy their desire for sexual intercourse with other women. Succumbing to such temptations can be destabilizing to marriages today because the extramarital partner is liable to be a candidate for marriage, or at least a stable relationship. In high sex ratio societies, such as most Arab countries today, there is often a double standard. Premarital chastity for women is emphasized and women who have sex outside marriage either are prostitutes in a technical sense or are regarded as women of ill repute. In either case, no respectable man would dream of marrying them. In double-standard societies, extramarital sex is not, therefore, a threat to marriage in the way that it is in the more egalitarian society of today in which extramarital sex partners can often be considered as potential spouses. In fact, in a more egalitarian society, women may sometimes deliberately engage in extramarital affairs as a means of alienating the affections of their lover from his spouse.

Even though women currently have the weaker hand in the marriage market, they often initiate divorce proceedings. The most frequent reason that women give for filing for divorce is that they feel a lack of emotional commitment from their husbands. The same reason is often given for initiating extramarital affairs, suggesting that this is one method used by women for extricating themselves from emotionally unsatisfying marriages.[15]

High current divorce rates reflect both the rising economic status and the declining birth rates of women. Since it is possible for women to have approximately the same income as men, the previously critical role of husbands in providing economic support for families has become far less important. Moreover, since having children contributes to the stability of marriage and since the birth rate is declining (as more women have gone into full-time jobs), this may be another reason that marriages have become unstable and the rate of divorce has risen.

Scarcity of men in America today compared to a century ago is due to many influences, such as greater male immigration then compared to now, greater vulnerability of men to accidents and diseases at all ages, and greater participation in wars. The death rate for younger women is much lower today than it was in the past when many young women died tragically in childbirth, often leaving several young children behind them. For this and other reasons, the life expectancy of women today is higher than that of men. Throughout history, most people did not make it to their fortieth birthday and women were less likely to do so than men. Agricultural communities of medieval Europe—particularly those in which people farmed difficult or marginal land—had high sex ratios, reflecting greater female mortality. This was partly due to their unsuitability for heavy agricultural work, which meant that male children received higher parental investment, including longer breast-feeding and preferential treatment during periods of food scarcity.[16]

The greater incidence of homosexuality among men compared to women—approximately 4 percent compared to 2 percent according to the most reliable scientific data[17]—provides a further drain on the supply of marriageable men. (It is not clear whether there is a higher percentage of homosexual men today or whether they are just more visible and accepted than they would have been a century ago.) Young women who hope for a better marriage market in the future should not hold their breath. According to U.S. Census projections, the current low number of men relative to women is going to persist almost unchanged for at least fifty years!

JEALOUSY AS A MOTIVE FOR DIVORCE

The central idea of an evolutionary explanation is that animals are perfectly suited in body and behavior to the way that they survive and reproduce. Hummingbirds have bills that are perfectly shaped to allow them to feed on the nectar from flowers; their wings, that move with a rapidity seen in no other birds, allow them to hover in flight while feeding. Just as you find an impressive lock-and-key match between the way an animal "makes its living" and the design of its body, so there is a fine-tuning of behavior and of the psychological mechanisms underlying human behavior. The intense sexual jealousy of men is an example of a psychological adaptation designed by natural selection.

Women have two basic requirements from the men in their lives. First, they want a good provider, a man who has reliable access to economic resources and is generous in contributing them to the family. (This is still desired, even by women who are themselves wealthy and powerful.) Secondly, the woman requires a man who is healthy and has a favorable genotype to pass along to the next generation. Physical attractiveness is an excellent proxy for genetic fitness, as already mentioned.

The dual aspect of a good husband as provider and mate creates a point of vulnerability for men. If a woman marries a man whom she finds sexually attractive and who is a good provider, everyone is happy. Suppose that the husband is personally a lot less appealing than his bank statement. This creates a temptation for the wife to divide up the duties of a husband between two men. Here's to our husbands and lovers, and may they never meet!

The husband-versus-lover dilemma is illustrated in the life story of Evelyn Nesbit, a beautiful young model who graced the covers of magazines with her innocent, ruby-lipped charm.[18] In 1901, at the age of sixteen she met and fell in love with Stanford White, a successful forty-eight-year-old architect. True to his reputation, White lost no time in plying the auburn-haired beauty with champagne and seducing her. Their subsequent frolics included Evelyn making use of a red-velvet-covered swing while stark-naked.

Much as Nesbit and White enjoyed each other's company, the young beauty realized that there was no future in their relationship. White was already married. He also entertained other women at the same time as his affair with Nesbit. She saw that it would be necessary to look elsewhere for a more permanent relationship, preferably with a wealthy man.

Her opportunity arrived in the person of Harry Thaw, the eccentric heir to a huge railroad fortune. Thaw made his case with blunt artistry by sending her a bunch of red roses wrapped in $50 bills. "Mad Harry" Thaw was famous for stunts like trying to ride his horse into the lobby of the genteel Union Club. Despite his eccentricities, Thaw turned out to be an attentive lover and Nesbit agreed to marry him.

Before their wedding, Nesbit made the mistake of revealing to Mad Harry the nature of her previous relationship with Stanford White. She was unaware that Thaw had a long-standing grudge against White, who had been instrumental in excluding him from some exclusive clubs. Consumed by jealous rage, Thaw soon afterward succumbed to the emotion by giving Evelyn a savage thrashing with a rawhide whip. Despite this trauma, she still consented to the marriage six months later, apparently because no better candidate for the role of husband had materialized.

Their troubles did not end with the wedding. If anything, Thaw's jealous rage got worse. Ultimately, he confronted White in Madison Square Garden's rooftop theater. White sat watching a musical. Thaw approached, fired three times, and killed him.

During the trial, which received all the attention of the O. J. Simpson case in more recent times, Evelyn saved the life of a husband she didn't love by revealing all the juicy details of her seduction by White. Following the trial, at which Thaw was found not guilty by reason of insanity and sent to an asylum for the criminally insane, Evelyn gave birth to a son whom she claimed was Thaw's (see fig. 9). Thaw claimed that it was White's.

Given their vulnerability to being cuckolded and wasting their reproductive investment on the offspring of another man, there is a selection pressure favoring wariness by men about the possible dalliances of their wives. Thaw may have been considered insane in his day, but his sexual jealousy differed from that of other men only in being more extreme. Men who were sexually jealous would have been more successful at preventing the extramarital affairs of their wives and would therefore have left more offspring. To the extent that jealousy is affected by genotype, and we now know that most personality traits are, the sons would have inherited their father's tendency to be suspiciously controlling of an attractive young wife.

Sexual jealousy has many tragic consequences, and it is a leading cause of violence against women. Such violence is punished severely in societies such as our own, but it is tolerated or even endorsed in societies of "honor," in which entire families are placed under a cloud by the adultery of one member. Acknowledging that the emotion has an evolutionary basis is obviously not saying that jealously motivated violence is justified. Men are presumably rational beings who can distinguish between right and wrong actions.

Fig. 9. Evelyn Nesbit with her son whose paternity was disputed by jealous husband "Mad" Harry Thaw who was convicted of murdering Nesbit's lover, celebrity architect Stanford White. (Reproduced from the Collections of the Library of Congress)

Both men and women may be highly jealous of a spouse, but there are interesting differences in the pattern of events that trigger jealousy, indicating that jealousy has been designed by natural selection to perform subtly different functions in men and women.

Women are more likely to become jealous if they feel that their husband is forming a deep emotional attachment to another woman, but are not as upset at the knowledge that their spouse is having a casual sexual relationship. Conversely, men are much more upset by the notion that their wives are having sex with another man than by the idea that she is becoming emotionally attached to the man. This adverse emotional reaction is as automatic and involuntary as the movement of a wound clockwork toy. It sets the stage for jealous behavior.

Jealous rage may be directed at the extramarital partner, or it may take the form of an attack on the spouse. When the husband or wife is killed, such attacks appear to be completely irrational; yet homicidal reactions are probably quite rare. The adaptive function of anger in the context of marital infidelity is presumably to control the partner and make them less likely to be unfaithful in future. Although women may experience the same intense jealousy as men, they are less likely to use physical aggression as a means of defending their marriage. They may, for example, work at improving their physical attractiveness, at denigrating and undermining their competitor, or at making home life more pleasant.[19]

Murderous rage is not the exclusive province of husbands, but it is more likely for them. This is as true in the age of the Internet as it was in the less sexually liberated world inhabited by Evelyn Nesbit and Stanford White. Take the case of Marlene Stumpff of Pottsdown, Pennsylvania, an unhappy homemaker whose search for love on the Internet led to her murder.[20] Pretending to be younger than her forty-seven years, she regularly exchanged flirtatious e-mails with a half-dozen men. It seems that she had a meeting with one of them in his home on the day that she died. Her new social life resulted in the running up of a $3,000 telephone bill that was a bone of contention with her husband.

Then, on January 20, 1997, Howard Eskin, a Philadelphia sports-radio talk-show host, sent her a dozen red roses with a card signed "Howard." This gift was discovered by her husband, which seems to have contributed to his rage. The couple began arguing as he did the dishes. According to police, Marlene struck her husband and he went wild. She was discovered by her son, lying in a pool of blood. She had been almost completely decapitated and three bloody kitchen knives lay on the ground next to her.

The idea that one's wife is being sexually unfaithful is deeply upsetting to most men. Many claim it is so disturbing they cannot bear to think about it, even for the purposes of a psychology experiment. When men do contemplate their wives' infidelity, there is an intensely unpleasant arousal of their peripheral nervous systems. For example, the heart speeds up as much as if the person had drunk two cups of coffee.[21] Hence, there is a violent response to infidelity evoked against adulterous couples in many countries—particularly "societies of honor" like traditional Greece, in which the jealousy response of the husband is extended to the woman's male kin, who are seen to be dishonored by her extramarital affair. The male proprietary attitude toward women is seen at a (possibly clinical) extreme in the case of wife-beaters and stalkers who may kill the object of their affection rather than contemplating her loss to someone else. It is no accident that women rarely stalk men, although many are vigilant for signs of infidelity in their husbands. Of course, stating that there is an evolutionary basis for the emotion of jealousy is not intended as a justification of any acts of violence that occur under its influence. Civilized existence is based on the ability of people to control their own actions, whatever they might be feeling.

SEXUAL JEALOUSY AND PATERNAL INVESTMENT

If men are not confident that children of a marriage are theirs, they may provide little support for them. In this way, they avoid investing in the children of another man and may form an emotional attachment to nieces and nephews. In many of the world's preindustrial societies, there is an especially close relationship between men and their sisters' children, an institution known as the *avunculate*. Men make a great fuss over nieces and nephews, while virtually ignoring the children of their own marriages. According to evolutionary biologist Richard Alexander, they do so because of a lack of confidence in their own paternity.[22] Men cannot reliably detect which children are theirs. If they could, the avunculate would not exist, and men would divert their paternal investment exclusively to their own children, refusing to provide for children sired by other men.

The avunculate can be explained in terms of genetic self-interest in a context in which men (a) have low confidence of paternity and (b) cannot detect which children are theirs. According to this view, genes that promote altruism will generally be successful only if the altruistic behavior is directed to close relatives who are likely to share the altruistic genes. By helping relatives, the altruist is increasing the probability that he will survive and reproduce, thereby increasing the number of copies of the altruistic gene in circulation. Men can promote their genetic interests by investing in the children of the marriage, or in nieces and nephews. Which of these strategies is preferred is largely determined by confidence of paternity or the probability that the child of a marriage is the husband's. Where men, on average, father only one-fourth of the children in a given marriage, it is in a man's better genetic interests to invest in nieces and nephews, because these would share at least 25 percent of his genes, and possibly much more in societies where marriage to cousins is practiced and the uncle is thus related to both parents.

In our society, husbands enjoy a high confidence of paternity, which means that it is very likely they are the fathers of children produced in the marriage. For this reason, American men invest in the children of their marriages, whereas men in societies having a lower confidence of paternity—such as the sexually more relaxed societies of the south Pacific—may divert most of their "paternal" investment to children of their sisters. Sexual jealousy not only produces violence against women, it also explains the pattern of paternal investment in different societies. Men who are sexually jealous of their wives have difficulty supporting the children of their marriage. In America, sexual infidelity is the single most common cause given by men in filing for divorce. By distancing themselves from the family, men are withdrawing their paternal investment even though they may be legally compelled to provide child support.

Seeing marriages as reproductive unions provides many clues as to why they are likely to fall apart. The strongest predictor of marital longevity around the world is the number of children. Couples with no children have a very high divorce rate that is ten times greater than the dissolution rate of couples with four or more children. Men are most likely to divorce when their wives are sexually unfaithful, whereas wives are most likely to become dissatisfied with marriages when husbands are emotionally distant or uninvolved. There is a natural life span for unstable marriages that varies around seven years. This evidently reflects the cooling of romantic passion. Love may begin cooling sooner than this. Still, the length of a typical unstable marriage would have been sufficient to raise a child to partial independence in a hunter-gatherer community.

Chapter Seven

Learning to Love

Our childhood experiences affect the kind of people we become. This is true even of parts of our adult lives that bear little direct relationship to childhood, including romance and sexuality. Natural selection has designed children to succeed in surviving and reproducing in the kind of social environments reflected in their home lives. Thus, children who have passed through an unhappy childhood, in which they suffered from a lack of emotional closeness and trust with adults, are "programmed" for a lack of trust and permanence in their love life. They often have a troubled romantic life ruled by intense but transitory infatuations. Psychologists who study the connection between motherly love and sexual love describe people who are happy in each of these relationships as being "securely attached" and, they see infant security as promoting security in adult relationships.

The insecure pattern is illustrated again and again in actors and entertainers whose marital instability is remarkable. Cynics feel that many of these marriages are little more than publicity stunts on the part of film studios, but it is implausible that independently wealthy performers would agree to turn their personal lives into a charade at the whim of some studio executive.

A recurring theme is childhood unhappiness. Popular entertainers like Marilyn Monroe and Billie Holiday are prime examples. Both had very unhappy childhoods. Monroe lived in numerous foster homes, in which she was exposed to emotional abuse. Holiday was raised in very difficult circumstances by a single mother who allowed her to become a child prostitute.

Monroe's romantic exploits are legendary. She married some of the most famous men of her day, including baseball star Joe DiMaggio and playwright Arthur Miller, and also created a scandal by her seductive singing at President Kennedy's birthday party; this fanned rumors that she was romantically involved with him.

Monroe's career as a screen siren really began in her early teens when she dressed up to excite the attention of passing motorists on her daily walk to school. From an early age, she was addicted to the social attention that her precocious sexual attractiveness received.

Billie Holiday's story is even more disturbing. She was attracted to abusive men who often beat her, as she freely admitted in the lyrics of her songs. Considered by some music critics to have had the greatest interpretive ability of any jazz singer, Holiday could reduce an audience to tears with the most banal of lyrics. Rumored to have been bisexual, she did not find much happiness in her romantic life. Like Marilyn Monroe, she sought relief through drug use. Both died early. Although the cause of Monroe's death remains mysterious, chemical dependency was likely a factor, as it clearly was in Holiday's case. If being in love produces a chemical high, as research suggests, then the constant need for romantic excitement can be thought of as a kind of chemical dependency. The highly public stories of these two stars are no different from the private struggles of many ordinary women.[1]

WHY PARENTS MATTER

In a recent book, *The Nurture Assumption*,[2] writer Judith Harris put forward the shocking idea that parents have little impact on how their children turn out as adults, and argued that children are socialized primarily by their peers. On the surface, Harris's book appears to be a compelling critique of much of the academic literature on the subject, but her argument was based on narrow reporting of the evidence instead of taking the entire body of scientific evidence into account. Parents do affect the social development of their children, as I concluded in *Why Parents Matter*.[3] Harris can only produce suggestive, or anecdotal, evidence in support of her alternative hypothesis that children are socialized by peers.

The influence of peers affects how children turn out, but not, as Harris believes, in the role of primary socialization agents. From an evolutionary perspective, there is an important reason why peers could not, in principle, have been the primary socializers of children. Peers are competitors. Being socialized by siblings and other playmates is rather like asking the KGB of old to train our CIA agents.

Evolutionary biologist Robert Trivers was the first scholar to point out that there is a strong conflict of interest between siblings in a family because they always want more parental investment than the other siblings want them to have.[4] The phenomenon of sibling rivalry is no mystery to anyone who has grown up with one or more siblings. Moreover, there is little doubt that this form of competition shapes our behavior, just as competition over status among peers influences the conduct of teenagers. Yet from an evolutionary perspective, these socialization experiences are of secondary importance to the primary role played by parents or other early caretakers in the formation of character and personality. Natural selection would have favored socialization by parents because they are the only people who can be relied upon to put the interests of the child first. In modern societies, we can see the harmful consequences when parents do not assume the primary responsibility for their children and allow them to spend a great deal of unsupervised time with peers. Children respond to what their peers are doing in order to become more effective competitors in the struggle for acceptance and respect, and this sometimes involves taking undue risks and breaking the law.

Although parents have the primary role in socializing offspring, we must recognize that parental behavior is shaped by social context. The adaptive logic behind this is that if you are going to live in a dangerous slum, you had better learn to look out for number one and avoid making the mistake of trusting others. On the other hand, if you live in a stable neighborhood where people are familiar with each other, it is more appropriate to trust people you meet. This might seem irrelevant to sexual behavior but it is not, because intimate relationships require more trust than most others.

ORPHANAGES AND EMOTIONAL DEVELOPMENT

Children raised in orphanages constitute a natural experiment that illuminates the importance of parents in promoting normal psychological and even neural development of children. Like all such experiments, it is far from being ideal as science. Some proportion of children in institutions may have had more developmental problems to begin with compared to noninstitutional children. For example, parents may be more likely to consign their mentally retarded infants to institutional care.

Orphanages of the past discouraged nurturant relationships between occupants and staff both as a matter of policy and of practicality, since there were often thirty children for each staff member. These policies have changed in the light of current information. Children that were raised from birth in the orphanages of the past experienced a variety of cognitive and emotional problems due to parental deprivation.[5]

Orphanages were much better at satisfying children's bodily needs than their social or psychological needs. Orphans received excellent nutrition and were properly clothed and bathed, but were deprived of adequate social stimulation.

Orphanage children sometimes experienced extreme social deprivation to control the spread of disease. In one institution studied in the 1950s by pediatricians Sally Provence and Rose Lipton, babies were isolated in cubicles for the first eight months to reduce the risk of spreading infectious disease.[6] (The identity of this orphanage was never revealed by the researchers.) Social contact occurred only when infants were fed or diapered. Feeding was accomplished by means of a propped-up bottle reminiscent of care for laboratory animals. Overworked attendants did not have the time to reassure each infant when she cried. Infants were not held, were not talked to, and were never played with.

Provence and Lipton compared the cognitive development of these orphans with that of children raised in families. During the first four months, few differences emerged. Soon afterward, the institutional children developed symptoms of maternal deprivation. The effects of social deprivation for infants were highlighted by a comparison between Teddy, an orphan raised in the institution, and Larry, a family-reared boy. Teddy showed the least developmental problems of his institutional peers and was deliberately chosen to present the most favorable picture of orphanage life.

Teddy's unusually good psychological development at the age of six months was explained by the researchers in terms of a peculiarity of his feeding arrangements. He had trouble maintaining contact with the nipple of his feeding bottle and would cry lustily whenever he had difficulty. This loud crying brought him more attention from the attendants. He therefore received more social stimulation than the other children.

Teddy's score on a measure of behavioral development (120 on the Viennese scale, a measure of early intelligence) was above normal and equal to that of Larry, the family-raised boy. At the age of six months, he was doing much better than the other institutional children who had pronounced developmental problems. Teddy's scores fell steadily, however. At eighteen months, he scored only around 65 on the Viennese whereas Larry maintained his original score of over 120.

Verbal descriptions of the institutional infants at the age of a year, before the steepest decline for Teddy, are revealing. The institutional infants rarely turned to an adult for pleasure, comfort, or help, and showed no signs of attachment to them. They could neither walk nor speak any words, milestones that most normal children have passed by this age. They were apathetic, showing little interest in the environment and little desire to play, either in solitude or with other children. They had disturbing symptoms such as constantly rocking back and forth in the manner now associated with autistic children. Behind the mask of solemnity, they expressed little emotion of any kind, seeming quite indifferent to pain as well as pleasure. Clearly, this would not augur well for successfully negotiating the complex interactions of lovers in later life.

Some of the institutional children were studied after they had been adopted into families. This is a more compelling type of natural experiment because it allows children to be compared with themselves, and thus gets around the problem of comparing apples with oranges that is encountered when institutional children are compared with those raised in ordinary homes. Most of the children adopted into homes in their first year showed huge developmental improvements after they had benefited from maternal attention and the social stimulation of family life. Great as these improvements were, institutional children had lingering problems with learning and problem solving, because the brain of an infant needs social stimulation from the beginning if it is to develop normally.

Their problems extended beyond academic difficulties to pervasive social problems. Institutional children had difficulty forming close friendships. They found it difficult to ask adults or other children for help. They did not have the rich imaginative life of ordinary children, as reflected in the simplified nature of their play. Social deprivation in the first year of life obviously compromises normal brain development.

Recovery of the infants adopted into homes demonstrated that there was nothing neurologically wrong with them at the outset. Maternal interaction early in life is essential for normal development of intelligence and sociability. Children who are not cognitively engaged by caregivers lose interest in their surroundings. Evidently, natural selection has programmed psychological development to occur normally against a backdrop of intense stimulation early in life from the mother and other adults.

The clearest evidence for the importance of social stimulation in normal brain development comes from the odd sensory deficits observed in well-known examples of socially isolated children, such as Victor of Aveyron, a French feral child whose story is accurately portrayed in the movie *L'Enfant Sauvage* (The wild child) directed by Francois Truffaut. Victor was cared for by teacher Jean Itard who was struck by his insensitivity to pain. Another example is Kaspar Hauser, who was kept in a stable from early childhood, apparently to remove him from contention as a royal heir, and was subsequently abandoned as a teenager in the streets of Nuremberg. He was deaf to some sounds and unusually sensitive to others. Amala and Kamala, the Indian wolf children, had difficulty seeing during the day. Abandoned by their parents, they had been nurtured by a wolf mother and were apparently most active at night when wolves leave their dens to search for food.[7]

Experiments on visual development in cats have shown that the feline's mature visual system is affected by the kind of light stimulation it receives early in life. Kittens that were allowed to experience only horizontal lines, for example, could not detect vertical barriers in a test maze and repeatedly blundered into them. Evidently the brain cells that normally detect vertical lines had ceased functioning. Neuroscientists have articulated a "use it or lose it" principle to account for such findings. This would clearly apply to the sensory problems of feral children. A similar explanation may apply to the language problems of socially isolated children, most of whom can master only a few words even after years of dedicated instruction. This also explains the problems of socially deprived institutional children.[8]

Sensory problems that are rooted in developmental brain anomalies are likely to be irreversible. Orphanages in which there is extreme social and sensory deprivation may produce sensory deficits resembling those found among feral children. Children adopted out of extremely neglectful Romanian orphanages during the 1990s had a persistently impaired capacity to respond to most types of sensory information, probably because their brains had not received enough stimulation early in life. They lacked the normal capacity to feel pain, hear sounds, and guide movements using visual feedback. Institutional children generally score about ten points lower in IQ tests than others, which is possibly due to the effects of early social and sensory deprivation on brain development.[9]

Enriched environments promote the development of brain cells leading to increased ability to solve problems. Social stimulation is a critical component of environmental enrichment. These phenomena were first demonstrated not among humans but among rats. Rats kept in group cages, instead of being housed alone, and provided with a variety of toys to manipulate became more "intelligent." They excelled in learning their way through complex mazes. The brains of enriched rats actually grew faster and brain cells established stronger connections (or synapses).[10] Enrichment effects are possible at any age among rodents, although the impact is greater early in life when the brain is more adaptable.

Children's brains are also highly receptive to stimulation early in life. Institutionalization, poverty, or having an inexperienced teenaged mother all reduce psychological stimulation of the brain in early life and decrease IQ scores. The importance of social stimulation and toys for children's intelligence has been demonstrated in Head Start enrichment programs. The early successes of Head Start programs were negated in the minds of many by the depressing tendency for early IQ gains to evaporate during adolescence. It is now clear, however, that if the enrichment programs are begun within the first two years of life—a sensitive period for neurological development—they can produce lasting gains in intelligence.[11]

The early environment is important not only for the development of intelligence but also for social skills. This aspect of being reared in an orphanage has received less attention from psychologists than the effects on intellectual development, but at least one study of Greek institutional children has examined the issue in detail. Researchers observed that children living in group homes were more demanding of teacher attention during class than children raised in families. On the playground, they stuck to playing with familiar children from the home because they had difficulty in forming new friendships. Relationships with other children from the home were far from amicable, however. There was a lot of fighting and the institutional children had trouble either trusting or confiding in each other. These results clearly contradict Judith Harris's claim that when children are deprived of interactions with parents, they manage just fine by resorting to peers.[12]

Being deprived of attachment to parents, and particularly to the mother, prevents the brain from receiving the kind of social information that it has been programmed to expect by a long history of evolution in which children always received social stimulation from their mothers and other adults. Children who do not have an opportunity to experience emotional closeness to others during childhood have difficulty feeling empathy for others and are likely to suffer conflict and loneliness in their intimate relationships.

SECURITY OF ATTACHMENT

With the advent of scientifically developed baby foods, there is no reason that men cannot raise infants, and many do. Yet in the evolutionary past, babies were fed on breast milk for several years and women did most of the child care. Moreover, there is a clear sex difference in evolved responses to infants that is reflected in the behavior and toy preferences of children. It is for these reasons, rather than any chauvinist antagonism to the economic advancement of women, that psychologists almost always refer to a young child's security of attachment to the mother. Children also become attached to their fathers and to other stable figures in their lives, including teachers.[13]

Introduced by developmental psychologist John Bowlby, the concept of *security of attachment* means that an infant is confident that the mother is there when needed. Bowlby was influenced by observing the physical attachment maintained between monkey mothers and infants. His student, Mary Ainsworth, developed a procedure for measuring security of attachment in humans in a laboratory setting that is known as the Strange Situation Test. In Ainsworth's test, reactions of an infant to temporary separation from and reunion with the mother are observed. Based on their reactions to the mother, infants can be categorized as belonging to one of several attachment types.[14]

If the baby wants to interact with the mother when she returns, he is classified as *securely attached*. He sees the mother as a dependable source of comfort and seeks the safety of her embrace. Infants who resist interaction with the mother after she returns are classified as *insecurely attached*. The mother is not perceived as a reliable source of comfort when they are upset. There are two kinds of insecure attachment: *avoidant* and *ambivalent*. The *avoidant* type of children ignore the mother and can be comforted by a stranger. The *ambivalent* children both seek and reject physical contact with the mother. They cry until they're picked up and then squirm restlessly until released. Two-thirds of infants are categorized as securely attached; most of the remainder are of the insecure avoidant type.[15]

The quality of the attachment between mother and child is influenced by the behavior of the mother, although some infants have more difficult temperaments than others and this affects the mother-child relationship. Although genes are important for most personality traits, they have no effect on attachment classification of a child.[16] Researchers have found that mothers of securely attached infants are more sensitive to the infant's needs than mothers of insecure infants. If the baby cries, for example, they respond quickly. They behave affectionately toward the infant. They are sensitive to the baby's unique preferences. They are more responsive to signals from the baby, using them, for example, to time the beginning and end of feeding sessions. Securely attached babies often express pleasure in their interactions with adults by smiling and gurgling.[17]

In many of the preindustrial societies, infants enjoyed a great deal of close contact with the mother, frequently spending the night in her bed and nursing on demand, conditions that would favor secure attachment. Children in modern societies usually sleep alone and are physically separated from mothers for much of the day. Despite this, most remain securely attached to her, suggesting that emotional closeness and sensitive responsiveness matter more than continuous physical closeness.

This intuition is supported by research on irritable infants that presents a special challenge to mothers because they may appear to cry, no matter what the mother does. Frustrated mothers may eventually give up trying to soothe the crying infant and attempt to ignore him as much as possible. Irritable infants seem to be that way for genetic, or temperamental, reasons. (Doctors used to refer to them as "colicky," but their problems may have little to do with gastric upset.) Thus, they are highly sensitive to unpleasant experiences and are continuously upset.

Not all irritable infants are insecurely attached, however, and an experiment conducted in Holland[18] shows that mothers can be trained to overcome the biological predisposition of their infants toward extreme irritability. In the experiment, mothers of irritable infants were trained to be more sensitive in responding to their offspring. Training occurred in three sessions when infants were between six and nine months old.

Mothers learned to understand their infants' signals and to react appropriately to them. When a baby becomes overstimulated, for example, it looks away. Instead of attempting to capture the infant's attention when this happened, mothers learned to wait until the baby showed signs that it was interested in social interaction again.

Infants of mothers who had been trained to be sensitively responsive were more than twice as likely to be securely attached by their first birthday (68 percent versus 28 percent) compared to irritable babies whose mothers had not been trained. Security of attachment is clearly affected by what the mother does. Similar conclusions can be drawn from research on abused infants who are insecurely attached to abusive mothers in 72 percent of cases. Many abused children grow up to be abusive parents themselves, although the majority do not.[19]

Early security of attachment affects many other facets of adult life. Securely attached infants become secure in their adult social relationships and are comfortable with intimacy in love and friendship. Women classified as insecure in their adult relationships have trouble maintaining long-term romantic attachments. They have many different sexual partners and are more likely to have babies as teens. Insecure men lack confidence and have difficulty forming romantic relationships. The sexual relationships of insecurely attached adults are also riddled with conflict—hence, the marital problems of film actresses like Marilyn Monroe who did not have a stable maternal figure early in life.[20]

EMOTIONAL CONFLICT IN CHILDHOOD

In *Death of a Salesman*, Arthur Miller, prior to marrying Monroe, dramatized the story of a man who made the mistake of investing everything in the pursuit of wealth to the detriment of family life. The story is a harrowing one because it highlights the fact that catering only to the material needs of a family, difficult though this may be, is not enough if their emotional needs are neglected. This scenario is relived when both parents work full time and the conflict between pursuit of wealth and emotional fulfillment in family life is even more pronounced. *The Death of a Saleswoman* remains to be written. Instead of being the story of a father's misguided sacrifice on the altar of Mammon, it would describe how children are sacrificed when parents are physically absent or emotionally unavailable due to the pressures of their occupations.

Emotional deprivation takes many different forms. Children may either lack adult role models or they may be exposed to coercive ones. For this reason, children who are raised by harsh abusive parental figures are different from those raised without parents in orphanages, or those suffering the milder neglect of parental absence during the day. Instead of being deprived of socializing experiences, they are provided with very clear role models. They learn, for example, that problems with other people may be solved through physical force. Since fathers are stronger than mothers, for example, masculinity may prevail in household disputes. In general, they reach the conclusions that the social world of adults is tough and competitive, and that people can be dishonest, cruel, and unpredictable.

Whereas orphanage children tend to be bland and inoffensive in their interactions with strangers, children raised in violent homes are more likely to be flamboyantly self-assertive and unwilling to do what is expected of them. They use hostility and aggression to control other people, both because this has been reinforced in the past and because they imitate the conduct of parents.

Research on the family dynamics of *conduct-disordered* (i.e., defiant) children provides a compelling example of how parental behavior is liable to get recycled in the next generation. Family members are in perpetual conflict with each other. Gerald Patterson, a psychologist and family therapist, describes such homes as "coercive" because one member of the family is constantly trying to force the others to do something, frequently to leave him or her alone.[21]

Parents in coercive homes rarely reinforce good behavior with physical displays of affection or with praise. Instead, they often attempt to suppress bad behavior using punishment. These efforts are frequently unsuccessful perhaps because the worst-behaved child gets the most notice and such social attention can be quite reinforcing, particularly if it is the only way that a child gets noticed.[22]

It would be very easy to dismiss the correlation between parental coercion and child misbehavior as due to the shared effect of "aggressive" genes. This view is incorrect, as demonstrated in successful interventions designed to reduce conduct disorders in children by changing parental behavior. Since parents can be taught more effective styles of interaction that reduce the level of defiance and aggression from children, we see that coercive parental behavior is at the root of the problem. Some of the most successful interventions with problem children have focused on the family as a social system.

Some clinical psychologists believe that children with conduct problems are produced by coercive family environments. For this reason, they like to work with the entire family. The objective is to modify the behavior of all family members so that the hostile tone of the household is reduced and ordinary friendly interactions become possible. Family therapist Gerald Patterson and his coworkers have produced good results largely by showing parents how to punish aggressive and coercive behavior of children (using time-outs and fines based on a token system), while rewarding desirable behavior with hugs and praise. These interventions produce good and lasting results, although the pace of improvement is slower in some families than in others.[23]

Interventions with difficult children illustrate the general principle that harsh and inconsistent discipline by parents effectively teaches children that there are no consistent rules worth following. If you cannot predict when a parent will reward or punish you, then it makes little sense to try pleasing the parent by obeying rules. Given that there is little justice, and your brother is always indulged despite his bad behavior while you are punished for being good, then you learn that looking out for yourself as the best policy to follow.

Being raised by harsh, inconsistent parents increases the risk of being delinquent, using illegal drugs, dropping out of high school, joining gangs, being unemployed, and becoming the parent of an unplanned child during the teen years (particularly for girls).[24] Children even come to treat themselves with the same lack of concern that their parents had for them. One practical consequence of this is a greater willingness to take risks in everyday activities such as driving, and a lack of interest in protecting health through wise lifestyle choices, including care in the selection of diet, avoiding tobacco use, and visiting a physician for checkups. Hence, this explains the facility with which many entertainers and others who have unhappy childhoods become addicted to alcohol or other drugs.

SOCIAL STATUS, PARENTAL INVESTMENT, AND LANGUAGE DEVELOPMENT

What we think of as social class, or social status, is tied up in most people's minds to having or not having money. Yet wealth is just one aspect of socioeconomic status. The social aspect is arguably just as important. One of the most important social facets of differences between social classes is the divergence in child-rearing practices. Affluent families have higher expectations for their children and, consistent with these high expectations, often but not always expend more personal effort as well as money in raising children, although money obviously cannot replace emotional closeness. This has important consequences for romantic development.

The critical role of parental behavior for children's emotional development is nicely illustrated in a small-scale but highly detailed study of interactions between parents and children from birth to the age of three years. Betty Hart and Todd Risley of the University of Kansas studied language development as a function of parental affluence.[25] They studied thirteen children of professional parents, twenty-three in working-class homes, and six in homes supported by welfare.

The most interesting results of the study provide detailed information on vocabulary development among children. The richness of a child's own speech was very much a reflection of what they heard from parents and was surprisingly uninfluenced by television and other sources of language stimulation.

Parents who were professionals used twice as many different words in conversations with their young children as welfare parents did. By the age of three years, children were articulating about half of the words used by parents to converse with them. Children of professionals uttered over 1,116 different words compared to only 525 words spoken by welfare children. Working-class children were intermediate, with a vocabulary of 749 words.

Children in professional homes received more quantity as well as more quality of verbal stimulation. Professional parents spoke an average of 487 words to their child each hour compared to 176 words by welfare parents. It takes two to have a conversation and children who had more speech directed at them were more verbal themselves. Children of professionals averaged 310 utterances per hour, almost twice the 168 produced by welfare children.

Although principally concerned with cognitive development, the Hart and Risley study constitutes one of the most compelling pieces of evidence to date for environmental influences on social development and provides some insight into why children raised in poverty are more likely to have difficulty in their future social relationships. The researchers took the important step of scoring the social comments of parents and children according to whether they were positive (i.e., warm or affirmative) in tone or negative (i.e., scoldings, prohibitions). The tone of the remarks addressed to children by parents was reflected in the children's own comments. Children hearing mostly positive conversation addressed to them responded with mainly positive conversation. Those hearing negative or rejecting comments from their parents responded in a similar vein.

The tone of parent-child relationships was much more coercive and unpleasant in welfare homes than in professional homes. What is really remarkable was the magnitude of the difference. When a parent spoke, a child in a welfare home was about five times as likely to receive a negative comment as a child in a professional home.

The emotional tone of parent-child relationships in homes of professionals is overwhelmingly positive. The emotional tone of parent-child relationships in the welfare homes was overwhelmingly negative. Welfare children heard twice as many rebukes as positive comments per hour (11:5). Children in homes of professionals, by contrast, heard six times as many positive as negative comments (32:5). Working-class homes were once again in the middle, suggesting that this was no fluke in the sense of the small number of welfare parents being atypical.

Just as children were busily imbibing the parental vocabulary, so they were acquiring a template, or pattern, for parent-child relations that would influence their own behavior as parents of the future. Even at the age of three years, children had already internalized the interactional styles of their parents. The researchers were struck that when they watched the children play at being parents of their dolls: "We seemed to hear their parents speaking. . . . We seemed to see the future of their own children." Of course, this is speculation, but it is informed speculation. In other words, children who were coerced by their parents used a negative tone in addressing their dolls. Children who were accustomed to hearing reasonable explanations about why they ought to behave in a certain way exercised the same sweet reasonableness in instructing their dolls. It is obvious that parental behavior is of overwhelming importance in providing children with their basic prototype of parent-child relationships. It is easy to dismiss such conclusions by arguing that what is transmitted is not a learned pattern for parent-child relationships but rather the genes for being a hostile, coercive person. Yet it makes no more sense to explain relationship patterns exclusively in terms of genes than it does to claim that vocabulary size is genetically encoded. A more reasonable interpretation is that our genotype gives all of us multiple potentialities whose realization depends on the rearing environment.

As already mentioned, studies of security of attachment for children and adults show that young children acquire not just a template for parent-child interactions but also a generalized model for interpersonal relationships. If the relationship with a parent is characterized by warmth and consideration, children will learn to expect the same in most of their future social interactions whether with friends, lovers, educators, colleagues, business contacts, or government officials.

All of this boils down to the conclusion that parent-child relationships tend to get transmitted across generations through a process of social learning. As an adaptive mechanism, this means that children are socialized using a style that helps them fit in with the social environment in which they find themselves. Children raised in welfare homes are being prepared to make their way in a difficult social environment characterized by a high probability of interpersonal conflict and economic difficulties (which is not to claim that parental conflict is absent from affluent homes). Children of welfare homes grow up to be suspicious of the motives of other people and this attitude affects their romantic lives, making it harder for them to achieve happiness in stable marital relationships.

THE IMPACT OF PARENTS ON ROMANCE

The Hart and Risley study suggests that children raised in poor homes are more likely to have a distant relationship with their parents, as reflected in the negative tone of conversations between parents and children. This lack of early emotional closeness has a profound effect on romantic relationships, particularly in the case of young women who suffer from a perceived deprivation of emotional closeness. This phenomenon can be illustrated by the case of young single mothers in inner cities.

According to Judith Musick,[26] single teenage mothers often feel deprived of love from their mothers, which makes it difficult for them to provide warmth and affection for their own infants. Paradoxically, their feeling that they have been deprived of intimacy makes them an easy prey for boyfriends who use them for sexual gratification and often abandon them as soon as they become pregnant.

These phenomena have been studied by sociologists in the case of young, single African American mothers living in inner-city slums. Black women find it difficult to marry, particularly those living in the inner cities, because of the scarcity of single black men with stable work histories. It is important to point out that there has been a mass exodus from such conditions by upwardly mobile African Americans who move to the suburbs.

When women are in a weak marriage market—i.e., they have trouble finding men with the personal and economic qualifications for marriage—they are often sexually active outside marriage. The availability of many young sexually active women means that men don't have to marry to be sexually fulfilled. Marriage actually closes off many possibilities for sexual relationships with various young women. This point is borne out by marriage statistics in the United States as a whole. During decades in this century when there was a particular scarcity of males, men married at a later age and delayed remarriage after divorce, as though "playing the field."[27]

Relations between inner-city black men and women are skillfully portrayed by urban anthropologist Elijah Anderson, who sees sexual relationships as a game of mutual exploitation in which women are likely to be the losers. Yet it is a mistake to see young, single African American mothers purely as victims. Some may decide to become pregnant because they want someone to love—a fantasy that may not survive the harsh reality of raising a child. The birth of a baby is often viewed as a positive social event in the life of a young black woman. She is welcomed into the baby club of other young women in a similar situation. Becoming a mother is a rite of passage that confers the status of adulthood and brings a feeling of authority and respect.[28]

Anderson describes the elements of the game through which "streetwise" young men seduce their female partners. For the young men, the object of the game is sexual gratification which they refer to as "hit and run" or "booty." Young women have dreams of romantic bliss of a more lasting kind. The game of seduction involves the man providing extravagant but dishonest promises of love, commitment, and even marriage. After the woman becomes pregnant, she is abandoned contemptuously, occasionally getting heaped with gratuitous insults, and the process begins with another woman. According to Anderson, the last thing that any streetwise male would do would be to take responsibility for raising the child he had fathered, assuming a high level of confidence that the child was his. Young men who do not have the stomach for this game of ruthless sexual exploitation are taunted as sissies and homosexuals.

According to Anderson, some of these young women do not intend to become pregnant outside marriage. Some use sex as a bargaining chip in seeking the attention and affections of a young man. Willingness to treat oneself as a sex object is characteristic of women who have received little affection in childhood; hence the traffic-stopping antics of the young Marilyn Monroe, who was thrilled by the power of her emerging sexuality to obtain the attention of passing male motorists as she walked to school.

Another striking feature of sexually provocative women is their weak reality testing. Most of the men they have met have been complete cads, yet they want to be swept off their feet by Mr. Right. Anderson's young women wanted to believe that if they became pregnant, their lover would marry them or at least provide some emotional support. They had a weak hand in negotiating with the opposite sex, however. Many had been brought up by single mothers themselves and had no father figure to exert pressure for a shotgun wedding.

Young women from emotionally difficult homes, such as those characteristic of extreme poverty, may be very romantic in the theatrical sense that we associate with actresses like Marilyn Monroe. As soon as one cad departs their lives, they are destined to meet the true love of their lives and fall even more desperately in love. Such romances can be exhilarating, but they also involve a pitch of excitement that is impossible to maintain over the decades of a stable marriage. The passionate love affair that has a tenuous hold on reality rarely lasts more than two or three years, and that is fairly typical for a Hollywood marriage. Lasting marriages are rooted in economic and social realities instead of being the flights of fantasy of romantic infatuations.

Chapter Eight

Sex
and
Work

Men and women have performed different types of useful work throughout the history of subsistence societies. Marriages were thus as much based on economic cooperation between the sexes as on sexuality. Men had to be good hunters to be desirable husbands, and women had to be efficient at gathering and good at caring for children to make satisfactory wives. Since the twentieth century, there has been considerable convergence of the economic roles of men and women that has affected the tone of modern marriages. Possibly for the first time in human history, it has become economically and practically possible for women to raise children by themselves; this has apparently weakened marriage ties.

Although there was a fair amount of variability in the sexual division of labor in preindustrial societies, men were mostly responsible for the killing of large game animals and women were mostly responsible for child care. These occupational specializations of our ancestors have resulted in persisting differences between men and women in brain specialization, in life goals, and in occupational choices. Anyone who wants to understand why men and women make different occupational choices, why there is a persistent sex gap in earnings in America despite almost equal pay for equal work, and why men continue to do far less than an equal share of housework ignores these evolved sex differences at their peril. To assert the reality of these differences is not to claim that there are some occupations for which men or women are biologically disqualified. There is no occupation in which men or women cannot succeed, if given the chance. The question is not whether men can be successful in raising children or whether women can succeed in climbing the top rungs of the corporate ladder. The real question is whether they *want* to.

SEX ROLES IN SUBSISTENCE SOCIETIES

The point of view that men and women today differ in their attitudes toward work and toward domestic duties partly because of inherited biological differences rests on the premises that (a) there was a fairly consistent sexual division of labor throughout human prehistory and (b) these consistent differences in work habits favored the emergence of sex differences in interests and cognitive abilities. Suggesting that there are average differences between the sexes to traits that are relevant to occupational choice and success does not minimize the importance of differences in training between the sexes. Average differences between the sexes do not mean that certain occupations are off-limits for one sex or the other. It merely provides a way of understanding why, given a fairly level playing field, there are so few women engineers and so few male elementary-school teachers.

We can be confident that sex differences in occupational choice today are not simply a byproduct of male political power and control over property. A study of subsistence activities in 224 preindustrial societies by anthropologist George P. Murdock, a leader in the construction of the Human Relations Area Files database, has shown that there is a distinct division of labor for almost all types of work.[1] Not only do individual societies divide up the work burden into male-dominated and female-dominated activities, but there is fairly good agreement between societies as to what constitutes men's work and women's work. Yet the cross-cultural evidence is very far from being a prescription about how men or women must behave because there is only one activity, metalworking, that is reported as an exclusively male activity. All other activities are performed by *both* men and women in at least one society.

Hunting large animals is a strongly male-dominated activity. In 166 societies, this is an exclusively male practice, whereas there is no society in which all of the hunting is carried out by women. Why might this be? It seems obvious that men have a number of bodily advantages and specialized cognitive abilities that make them likely to be more successful hunters of large animals. They include:

- Greater size, running speed, and upper-body strength.
- Faster speed of reaction.
- Better throwing accuracy, making a damaging strike at a prey animal with a spear or other projectile more likely.
- Better three-dimensional spatial ability, which would be important for planning and executing a cooperative hunt.
- Better navigational skills which would help hunters travel efficiently to and from hunting grounds.

There is a very simple explanation why there is no society in which hunting is the exclusive duty of women. Men specialize in this work because they have adaptations that make them better hunters. The identification of hunting as a masculine specialization reflects the pragmatic desire of people in subsistence societies to improve their chances of eating meat. Even in this sphere of clear masculine advantage, women may play an important role. In thirteen societies, both women and men engaged in hunting large animals together. A similar pattern applies to the trapping and catching of small animals, except that there are two societies in which sex roles are reversed, and trapping is an exclusively feminine activity.

The gathering of food of various kinds (fruits, berries, nuts, herbs, roots, seeds) is almost seven times as likely to be an exclusively female activity as an exclusively male one. In approximately one-quarter of the societies, it was practiced by both men and women. Gathering may not be as specialized an activity as hunting, but it is hard to avoid the impression that women monopolized gathering because they were particularly good at it. Some of the cognitive abilities relevant to gathering include:

- Women's much better manual dexterity that would allow them to harvest small fruits and berries more rapidly.
- Women's faster recognition of complex visual patterns that would allow them to identify different food items.
- Women's better color sense (or at least rarity of color blindness) that would allow fruits of exactly the correct ripeness to be selected.
- Women's better memory for the location of objects would allow them to harvest all of the ripe fruits on a bush more rapidly.

While it is reasonable to assume that sex differences in cognitive abilities—arising from a history of natural selection for different subsistence activities—have played a role in the division of labor among societies studied by anthropologists, this is certainly not the whole story. Thus the fact that women are most likely to be responsible for domestic chores such as cooking, hauling firewood, basket making, manufacture of clothing, carrying water, and making flour, is probably a side effect of the fact that they spent more time close to home. There are some societies in which these activities are performed exclusively by men, however.

One fascinating aspect of the cross-cultural research is that the subsistence activities of preindustrial societies were not more sex-typed than our own, but in some ways, less so. For example, there are far more male than female farmers in America today. Yet in subsistence societies, agricultural work was more evenly divided between the sexes. Women did much of the work of tending crops, whereas men were more likely to do the heavy work of clearing trees in preparation for planting. What has changed is the increasing economic power of women that has made them far less dependent on the role of their male partners as providers than they were in the traditional agricultural societies of the past. Even if women do not have exactly the same jobs as men, they approach males in earning power, and the fact that women still earn less than men in economically developed countries is partly a function of their occupational choices, although sociologists often emphasize the diminished economic choices available to people at the bottom of the economic ladder who may feel obliged to accept any work that is available. Sex differences in work roles still prevail in two spheres that were not even examined in Murdock's study—namely, which sex takes care of, feeds, and socializes children, and which sex wages war.

THE PERSISTENCE OF OCCUPATIONAL DIFFERENCES BETWEEN MEN AND WOMEN

The generalization that different societies around the world carve up the workload into duties that are considered "men's work" and duties that belong to women still applies in the United States, despite equal opportunity legislation. This is true to an astonishing degree, considering that men and women are educated similarly, that there is legally protected equality of opportunity for men and women, and that there is widespread agreement among both sexes that women should be represented in all occupations—including those that were exclusively male in recent history. The only serious reservation that is publicly expressed on the matter is whether women should serve as combat soldiers.

The distribution of men and women in many occupations today is far from the roughly fifty/fifty split predicted by the numbers of men and women, and this has several possible explanations apart from evolved sex differences in aptitudes and motivation. Young women may not even consider being lumberjacks because they see this as a stereotypically masculine occupation. Similarly, many men may not give a second thought to the possibility of becoming a cosmetologist. They can be said to lack role models for these occupations. Yet the lack of role models is not such a crippling problem as might be imagined. During World War II, large numbers of American women went to work in factories for the first time to replace men who were abroad fighting (see fig. 10).

Such women were referred to as "Rosies," after Rosie the Riveter. "Rosie the Riveter" was the name of a popular song and the woman described in it was fictitious. Then in 1949, Walter Pidgeon discovered Rose Monroe, working as a riveter at the Willow Run Aircraft Factory in Ypsilanti, Michigan. The Hollywood star was making a film to plug war bonds, and instantly realized the publicity value of a real-life Rosie the Riveter—he signed Rose Monroe for his film.[2]

Rose Monroe was a trailblazing woman. She went on to found Rose Builders, a home construction company that operated from Clarksville, Indiana. She also had a passion for flying, and earned her pilot's license. A fascinating aspect of this story is that after the war, other Rosies did not continue in their masculine occupations, as their poster girl Rose Monroe did. Many simply went back to being homemakers and let men do the metal fabrication. Whatever else we might conclude from this, it seems clear that gender plays an impor-

Fig. 10. During World War II, many women took on jobs such as working in aircraft factories, positions previously held only by men. (Reproduced from the Collection of the Library of Congress, LC-USZ62-112283)

tant role in occupational choice even when the issue of role models is somewhat neutralized by actual job experience. It seems that the Rosies were happy to have supported the war effort but sought fulfillment in traditional roles as wives and mothers. Of course, the prevailing belief at that time was that men were the main providers and that women took on these roles only under extraordinary circumstances, such as a wartime emergency. Since then, the institution of marriage has weakened—as reflected in rising divorce rates, for example—and women have become more career-oriented.

Another important issue in occupational choice is pay. In general, occupations that are perceived as feminine tend to be more poorly paid than those that are masculine, if education level is equal. Thus, elementary-school teachers earn less than computer programmers. While it is easy to dismiss this situation as economic discrimination against women, no one is being forced to be a teacher rather than a programmer. The decision to take up elementary-school teaching as an occupation is one in which the benefits of the job must outweigh its drawbacks. There is some evidence that men are more motivated by salary in their choice of occupation than women are. Thus, cross-cultural research in twenty countries found that men valued money more than women did.[3] Such gender differences in preferred job attributes are generally quite small and have therefore evoked some controversy among scholars. For that reason, the results of 242 different studies were combined in a meta-analysis. Based on the combined body of research, small sex differences were found that were consistent with evolutionary predictions about sex-specific adaptations. Thus, men valued promotions, leadership, power, and challenge more highly than women and women valued interpersonal relationships at work, opportunities to help others, and intrinsic aspects of their jobs more highly than men.[4]

The sex differences in preferred job attributes helps to explain why so many more women than men are employed as teachers. Men do not reject teaching as an occupation. They are perfectly happy to teach for a living if the salary is high enough. Thus, there are still far more men than women teaching at the college level, despite fairly determined efforts by college administrators to hire women in response to political pressures both subtle and overt. Still, the men in this field have a good head start and there are only a finite number of positions leading to tenure.

In addition to a greater focus on money in occupational choice, men are more career-oriented in other ways. Among married couples with children, true dual-career families are rare because it is extremely difficult for a couple to manage two careers in addition to the responsibilities of raising children. In these cases, it is unusual for men to give up their careers to raise children. A far more common pattern is for the wife to work part time, often in a dead-end job, while taking on most of the responsibility of caring for children and managing the home. One recent report found that only about one woman in six succeeds in combining a career (defined as income above one-quarter of Americans) with having children, suggesting that the work of raising children still forecloses career opportunities for many women.[5]

Men are not only more likely to have careers, they also have more energy to invest in them because they are less committed to child care and housework, although men are contributing more time than in the past to child care. One useful measure of career effort in academic life is the number of professional publications that are turned out. Woman's academic work may be of the same quality as that produced by men, but their publication rate is lower. This implies that when they are confronted with a conflict between professional development and family life, academic women may assign a higher priority to the competing demands of home and family than men do. It is important to realize that these differences are differences in general: individual academic women may be as strongly career-motivated as individual men, winning Nobel Prizes, for example—and the same is true of every other profession, including those in the corporate world.[6]

A Complex Look at the Glass Ceiling

The decision as to whether the husband or wife pursues a career in the case of married couples with children has often been attributed to societal influences, such as economic discrimination against women by employers, patriarchal attitudes in the home, media stereotypes, and so forth—but these possible explanations should not blind us to the importance of individual choices by men and women. Evolutionary psychology illuminates the issue of why women are more influenced by the desire to care for children, whereas men are more influenced by the desire to establish careers. These are unpopular phenomena in the current climate of sexual egalitarianism but merely disliking them does not make them go away. Thus, most men probably believe that their wives should have every opportunity for career advancement as long as they do not have to sacrifice their own careers to care for children, or do an equal share of housework when their spouse works full time.

Although women constitute some 40 percent of the U.S. workforce, they hold only about 6 percent of senior executive positions. This phenomenon has been described in terms of a "glass ceiling." The glass ceiling is a metaphor suggesting that while women may rise to a certain level, there is an invisible barrier within corporations that prevents them from getting into the boardrooms of Fortune 500 companies. From asserting that there is such a force and that it is external to the women executives, it is a small step to assigning blame within the corporate hierarchy and arguing that the inequity is due to discrimination against women in promotional practices, but the truth is more complex than this.[7]

If there is discrimination against women in corporate America, it is of a very peculiar and selective kind. Companies have no trouble hiring female executives. In fact 40 percent of executives are women, exactly what would be predicted from their representation in the workforce. When one Fortune 500 company was sued for not hiring enough female managers, it commissioned a study to determine the reasons. The study found that differences in promotion rates reflected sex differences in attitudes and behavior. Women were much less willing to relocate for promotion, for example, or to work longer hours, perhaps because of competing social obligations. They were also less inclined to see their current job as a stepping stone to higher positions.

These sex differences may be a complex result of contextual factors. For example, women are probably less willing to relocate because they value their social network more highly than men do and are therefore more unwilling to move to a city where they know no one. Unwillingness to work longer hours could, in theory, be affected by more of a commitment to being with children in the evenings. Yet the study found the same sex difference for women who had no children. Among married people, there was an interesting sex difference. Marriage increased men's efforts to obtain promotion, but it decreased women's desire for promotion. The study concluded that the lower promotion rates for women were not caused by discrimination, but by their own attitudes and behavior. Women who looked for and accepted responsibility were promoted at the same rate as men. It should be noted, however, that this was the result of just one study.

Although it is difficult to establish objectively that most corporations actively discriminate against women in their hiring practices, studies of women who have risen to the level of corporate presidents and chief executive officers reveal a peculiar dynamic of social interactions that can be particularly uncomfortable for women.[8] If a corporate executive does anything that allows her to be perceived as a sex object, she loses credibility. (An underlying logic for this is suggested in the next chapter.) Dress and hairstyle must always be formal and sober. One executive refused to play tennis with her colleagues because she felt it was impossible to be taken seriously in tennis attire. The same restrictions on behavior help women to fit in with the "old boys' network" that comprises corporate America. A woman chief executive does not expect men to open the door for her. If she is to be accepted as an equal, she must behave at times more like a man, even if this means cultivating an interest in sports and other masculine topics of conversation. Similar sex-role pressures are experienced by women at lower levels of corporations, but the difference is that they spend more of their workday interacting with other women.

The glass ceiling is nevertheless a misleading metaphor because it attributes wholly to external forces, such as discrimination, outcomes that may be in part internal and reflect evolved sex differences. Studies of people who rise to the top in organizations have shown that they have stereotypically masculine traits, whether they are male or female. They are assertive, competitive, and domineering. They refuse to buckle under pressure, and they are willing to take risks. Women who have these "masculine" traits are more successful in their careers. By contrast, women with stereotypically feminine traits, such as nurturance, accommodation, and a desire to soothe hurt feelings, do less well in business careers.

The greater male investment in careers reflects an evolutionary past in which competitiveness was rewarded by increased reproductive success for men, but not for women. Money is more important to men in their choice of a career than it is for women. In return for higher wages, men are willing to take on risky occupations, which helps explain why over 90 percent of accidental workplace deaths are male. Moreover, men are greatly overrepresented in jobs that most people see as highly unpleasant. Men constitute 95 to 100 percent of employees in the twenty-four most disagreeable occupations in terms of work environment, stress, salary, job security, future prospects, and physical demands. This suggests that men are willing, or at least expected, to make more of a sacrifice to earn a living than women are.

The fact that American women earn about seventy-five cents for every dollar earned by men has often been used to suggest that they are treated unfairly by employers, but such statistics make little sense without taking a number of other sex differences into consideration. In America, women with children work shorter hours than their husbands do. Women are more likely to leave the workforce for short periods than men are, which tends to reduce their earning potential. Men are also more driven by the need for financial success, whereas women are more willing to sacrifice salary considerations for better working conditions, flexible time, a shorter commuting distance to work, and the opportunity to help others, in addition to interpersonal aspects of the job. This doesn't mean that the employers should exploit them. Women are more attracted to public service jobs than men are, possibly because these jobs provide an opportunity to help people and perhaps because they generally provide good working conditions and job security. Taken together, these various sex differences help to account for the persistence of pay differentials between men and women—but again, this does not deny that employers have sometimes paid women less because they thought they could get away with it, which has exposed them to costly and embarrassing lawsuits.

SEX DIFFERENCES IN SPECIFIC OCCUPATIONS

If men and women retain the freedom to choose their occupations, the ideal of a world in which all occupations are equally divided between men and women is improbable for two reasons. The first is that men and women have average differences in cognitive abilities that affect occupational choice at least in some fields. The second is that men and women differ in evolved motives. Their choice of occupation is affected by the desire to satisfy different psychological needs.

Attributing occupational choices of men and women solely to sex differences is very risky, however, because of the slew of other potential influences. For example, there were no female soldiers or police officers a hundred years ago because women were not allowed to serve in these capacities, much as they might wish to do so. This led to a whole genre of folk songs in which women cut their hair and go off to war, or to sea, disguised as men. It is also risky to predict the future of people's occupational choices. One way of getting around these problems is to study specific sex differences in particular occupations. If these are very large, and cannot be explained by restrictive legal and economic influences, it is reasonable to suspect that they may reflect fairly stable sex differences in aptitudes and interests. If the differences persist virtually unchanged despite decades of legally enforced equality of opportunity, then this supports the view that evolved average differences between men and women are playing a role.

The U.S. Census Bureau collects information on occupations and this is organized in a useful table in the *Statistical Abstract* that compares female representation in various occupations in 1983 to that in 1997.[9] This is an interesting period to look at because it has seen the entry of unprecedented numbers of women into the workforce at a time when barriers to women were being removed by the enforcement of laws favoring equal opportunities by sex. If rapid changes in the representation of women in various male-dominated occupations were to occur, they should have begun in this period.

In fact, there have been some changes, but the most substantial shifts have occurred in government jobs which are particularly susceptible to affirmative action policies. The U.S. Postal Service is a good example. Over the fourteen-year period, there has been a large increase in the number of women hired by the Postal Service. For example, the number of female mail carriers has almost doubled, from 17 percent in 1983 to 31 percent in 1997. Interestingly, the number of female messengers—who perform virtually the same function in private enterprise but usually for less pay—declined from 26 percent to 23 percent in the same period.

There has also been a large increase in the number of female professional athletes. This used to be a stereotypically masculine occupation, but there have been determined efforts to promote female sports participation in schools and colleges. These efforts have probably made a difference both in promoting public interest in women's sports and in encouraging women to make a living in them. Yet most sports remain segregated by gender. There are very few mixed sports, and there are few professional sports in which women compete as equals with men.

Being active in sports used to be associated with a masculine identity, but this is no longer the case. Women today identify more strongly with the ideal of physical fitness than they used to. This interest is strongly associated with increased entry of women into the workforce, which has produced a shift toward a more slender standard of attractiveness that is promoted by physical activity as well as dieting (see next chapter).

Education is another interesting profession from the point of view of changing participation of women. As noted already, it is mainly women's work at the elementary level where pay and occupational prestige are low. College and university teachers, who receive the highest salaries in the profession, are predominantly male. In recent decades there has been a determined effort to employ more women, at least in entry-level positions, as part of a political agenda for colleges and universities which must live up to Equal Opportunity employment practices to be competitive for public funding. Even with aggressive affirmative action, which means hiring women in preference to better-qualified male candidates, there has been a comparatively modest increase in the proportion of women teachers at the third level, from 36 percent to 43 percent.

At the same time, women have actually increased their representation at lower levels of the educational system, from 71 percent in 1983 to 76 percent in 1997. This means that sexual segregation of teaching has *increased* at the lower levels. All sorts of teaching requires an element of nurturance, and this takes a form that is closest to stereotypical maternal behavior for younger children. Consistent with this impression, the proportion of female kindergarten teachers has remained steady at a whopping 98 percent. Moreover, the proportion of women speech therapists, whose work is primarily with children, has increased from 91 percent to 95 percent. The same pattern is found for most other occupations involving care of children. The proportion of female childcare workers has scarcely budged from 96 percent to 95 percent.

Another profession that has been historically feminine and based on a capacity for nurturance is nursing. The numbers suggest that nursing is not about to be sexually desegregated in the near future. The proportion of female registered nurses has declined from 96 percent to 94 percent, a difference that is too small to notice in the real world. Similarly, the proportion of female dental hygienists has scarcely changed, from 99 percent to 98 percent. Apparently men going into the medical field still do not consider nursing or other support occupations, possibly because they, like male educators, aim at the better paid occupations in the field. Women, however, have also entered these more lucrative fields in large numbers. The proportion of female physicians has increased from 16 percent to 26 percent and the proportion of female dentists has increased from 7 percent to 17 percent. These increases suggest that, in time, there may be as many female as male physicians and dentists.

Female progress has been slower in engineering. Engineering requires very good mathematical and spatial abilities. Even though the average difference between the sexes in mathematical abilities is vanishingly small, cognitive abilities specialists have concluded that men are greatly overrepresented in the range of people with exceptionally good mathematical and spatial abilities, which helps them gain entry into engineering schools.[10] If these views are correct, then it would be predicted that (a) there are very few female engineers and (b) the number of female engineers will not increase greatly over time.

In 1983, only 6 percent of engineers were women. By 1997, this had increased to 10 percent. This is not a large *absolute* increase but it is a large *relative* one. A difference this size could be due, in part, to affirmative action policies at engineering schools. It remains to be seen whether large numbers of women will cross the mathematics hurdle. Cognitive abilities specialists suggest that they may not. Engineering could remain a male-dominated occupation both because women are less likely to gain entry to engineering schools and also, perhaps, because they see the profession as "dorky" and unfeminine.

One of the most male-dominated occupations is that of auto mechanic. This is hardly too surprising when you consider that there is a relevant stereotypical sex difference in toy choice. Boys are much more likely to choose vehicles as toys. The reversed toy preferences in the case of CAH (congenital adrenal hyperplasia) women, whose brains are masculinized by exposure to testosterone during a critical period before birth, strongly suggests that this interest is biologically determined.[11] In absolute terms, the increase in numbers of women serving as auto mechanics has been small, from 1 percent to 2 percent. Being a mechanic is obviously closely related to the development of interest in vehicles, which is much more of a masculine interest from childhood; this may be related to the visual–spatial adaptations for hunting discussed in the first chapter. Given that this interest is unlikely to change, it is probable that auto mechanics of the future will continue to be predominantly male. Women may also steer clear of the occupation because it is dirty and women are underrepresented in such physically unpleasant jobs. It is interesting that far fewer women than men work as drivers even though most women drive. In 1983, 9 percent of drivers were women and in 1997 this had increased to only 11 percent. Some of these jobs, such as being a taxi driver, are dangerous and so may not appeal to women.

There has been little change over time in the types of office jobs that have traditionally been held by women. Men may have been prevented from entering these occupations, both by the fact that salaries have been low and the fact that the position seemed to fit a feminine stereotype. In 1983, no fewer than 99 percent of secretaries were women and in 1997 it still stood at the astonishing figure of 99 percent. The proportion of female typists slipped between these years, but only from 96 percent to 94 percent; yet with the universal adoption of the personal computer, these occupations have declined in numbers and will likely continue to decline.

Why are there so many female typists? One reason may well be that typing as a service occupation has little career potential, so men don't want it. Yet the same is true of food preparation and service industry positions that were 37 percent male in 1983 and have increased to 41 percent male. It is hard to evade the conclusion that women gravitate toward typing as an occupation because it provides good working conditions, often with flexible hours, and because they are particularly good at it. Typing calls for fine manual dexterity in which the fingers have to be separated and aimed precisely at a target at high speed. Formal tests of manual dexterity—such as the pegboard test and being asked to move fingers individually—indicate that women have a clear advantage. It would be surprising if they did not also demonstrate this skill in typing. This is certainly not the only reason that women have been typists and secretaries. Women who lack education and need flexible hours would gravitate toward these occupations despite their low social status. The proportion of female receptionists has held steady at 97 percent suggesting that women are either much better at this work than men—by virtue of superior verbal and interpersonal skills—or that they find it rewarding or lack education and training for better-paying positions. Another factor may be that receptionists are useful, with their physical attractiveness, as friendly greeters in front offices.

A sphere in which women may have an advantage is in the field of personal relations, including human resource management and public relations. Many women have highly responsible positions in the field of public relations. In formal public relations occupations the proportion of women has increased substantially from 50 percent to 66 percent, suggesting that they are about to establish clear predominance in this profession. To the extent that this profession is based on interpersonal communication, it might provide an opportunity to express female social adaptations. According to social psychologist Sharon Brehm, of the University of Kansas, "Compared with men, women are more skilled at sending and reading nonverbal messages, listening to others, and consoling those who are distressed."[12]

The proportion of women in service occupations as a broad category—including cleaners, waiters, police, nursing aides, and hairdressers—has remained substantially higher than the proportion of men, falling slightly from 60 percent to 59 percent. Women continue to hold most jobs as cleaners and servants, just as they do most of the domestic work in societies studied by anthropologists. The proportion of females in these occupations has held steady at 97 percent. Many women in these occupations are immigrants who often lack educational credentials and may not speak English. However, the number of female police officers, who are better paid, has increased from 6 percent to 12 percent.

In one of the changes that has received much attention, women have made strides in executive, administrative, and managerial positions, increasing their share of management jobs from 32 percent to 44 percent, which suggests that they are about to achieve parity, or better, with men, in the near future, even if they are underrepresented in the top management of large corporations. Females have maintained a high representation in sales occupations, increasing their proportion from 48 percent in 1983 to 50 percent in 1997. Women seem particularly well suited for careers in business, perhaps because skill in interpersonal relations is so important in this sphere.

In general, there are large and persistent sex differences in many occupations. This is not to say that either men or women are disqualified by gender from any occupation they want to enter. Evolved differences in aptitudes and interests clearly underlie some of the substantial sex differences in occupational choice. Many of these sex differences are relevant to the division of labor in preindustrial societies and arguably reflect complementarity of roles among our hunter-gatherer ancestors. They are relevant to the romantic tie between the sexes because marriage has always been, among other things, an economic union. Evidence of such ancestral complementarity can also be seen in respect to the care of children. Women's reproductive roles were much more important in the past than they are today but, as discussed in the next chapter, women are more likely than men to be evaluated in terms of reproductive stereotypes; this has important implications for how they present themselves, and how they are perceived, in careers.

SEX DIFFERENCES IN CHILDCARE AND HOUSEWORK

The persistence of many strong occupational sex differences—despite political and social changes in favor of equality of opportunity—is mirrored by a curious persistence of sex differences of domestic work done in the home. It is true that the increasingly important economic contribution of American women to their households has meant that men today are doing more domestic work than ever before. Yet even when women contribute as much economic support to the family as their husbands, they continue to do far more of the routine domestic work like cleaning and laundry, and they take far more responsibility for the care of children.

Although many social commentators attribute inequality of work responsibility to a relative lack of social power on the part of women, a more balanced interpretation is that women are more strongly motivated to care for children and homes than men are. While American fathers perform more domestic work today than they did in the past, cross-national comparisons suggest that their participation in care of children, for example, is similar to that of other fathers around the world. In one thorough study, John and Beatrice Whiting,[13] of Harvard University, compared the amount of time children spent in the presence of or engaged in activities with their mothers compared to their fathers. In the United States, mothers did five times as much of the childcare as fathers. In India, mothers did more than twelve times as much as the fathers. For the other countries studied by the Whitings (Kenya, Mexico, the Philippines, and Japan), mothers took on three or four times as much of the childcare duties as did fathers.

One interesting sidelight on these findings is that they are not simply attributable to an inability on the part of fathers to care for children. In the case of custodial fathers in the United States, we know that men can be just as effective as women in raising children, with some minor qualifications. In fact, when men care for small children, they interact with them in ways that are strikingly similar to maternal behaviors, even switching to baby talk, for example. Developmental psychologists have also concluded that men can provide competent routine care of infants. Even though fathers may be effective care givers, it cannot be denied that there are stereotypical sex differences in parenting style. Fathers tend to engage in more rough play with children, for example. They are also less diligent at monitoring their children's moment-to-moment activities.[14]

While the lower average participation by men around the world in care of children might be due to greater absence from the home at work, this is certainly not the whole story. When both parents are present in the home, mothers are almost twice as likely to interact with the children, and they provide more than three times as much routine care as fathers. These results were produced in studies conducted in the United States, Sweden, and Israel.[15]

The amount of domestic work performed by men reflects their economic power and social status. Among the Aka pygmies of central Africa, low-status men apparently compensate for their lack of prestige by an unusual willingness to care for children. It has been estimated that Aka fathers provide more direct care to young children than fathers in any other society. When studied in camp, Aka fathers held their infants more than one-fifth of the time. Yet over the course of an entire day, they held the infant far less than the mother did.[16]

In America and other industrialized countries, the increasing economic power of women has meant that men do more housework. Among Australians, for example, male domestic work increased from twelve hours per week in 1974 to eighteen hours in 1992. This is still dwarfed by the thirty-four hours per week performed by women. A similar pattern has been seen in other industrialized countries.[17]

Even if men are doing far more domestic work than they used to, they are still doing far less than their wives—even when both partners work. A recent study of American families in which both partners worked full time found that women still did twice as much of the housework as men. The fact that women have to put in long hours at work for pay and also work a "second shift" at home is sometimes referred to as exploitation. Yet the exploitation view is only part of the story.

Research indicates that the attitudes of wives are a more important predictor of how much housework husbands do than the attitudes of the husbands themselves. Thus, husbands may believe in equality of roles of husbands and wives as far as the performance of domestic chores is concerned, but they may actually do very little domestic work. Men make a substantial contribution to housework only if their wives also have liberal attitudes and believe that men should do much of the housework.[18] Working women may continue to do all of the housework because some are unwilling to relinquish control over the domestic scene. For whatever reason, there is an interesting persistence into modern times of the cross-cultural trend of women doing most of the domestic chores. They also continue to do most of the childcare. The stronger motivation of mothers to care for children is likely part of an evolutionary heritage in which women spent much more time looking after children. This argument does not imply that men cannot take care of infants, or that they are not motivated to do so, but merely that the motivation of mothers is generally stronger.

MATERNAL CARE AS AN EVOLVED BEHAVIOR

The fact that women in all societies are much more active in caring for their children and teaching them social skills than men are suggests that they are more interested in doing this. It is not simply a question of being exploited, or being left holding the baby literally and metaphorically. Study of the behavior of noncustodial parents following divorce is quite revealing here. Noncustodial mothers typically remain in daily contact with their children following the divorce. They are motivated to visit the child on a regular basis, have the child stay with them overnight, and keep up contact through frequent letters and phone calls. By contrast, only one child out of six enjoys regular contact with the noncustodial father. Most do not even see the father from one end of the year to the next. Divorce thus causes a marked deterioration in the relationship between children and their noncustodial fathers, but has little effect on the relationship with the noncustodial mother. This difference is evidently due to the greater emotional commitment of mothers to stay in touch with their children. It cannot be explained away as following some arbitrary social role that dictates how mothers should behave. For one thing, denial of custody gives mothers a legitimate reason for staying away if they wanted to evade the responsibility of their children.[19]

Mothers evidently become more deeply attached to their children than fathers do. The underlying mechanisms have not been so well worked out. One possibility is that the very different quality of the maternal relationship could partly reflect the biological realities of pregnancy, childbirth, and breast-feeding. It seems almost inevitable that the emotional quality of these experiences should affect the subsequent relationship between mothers and children.

Moreover, the greater interest of young girls than young boys in caring for infants is possibly due to hormonal events early in development that produce sexual differentiation of the brain. Research has even shown that the hormones circulating in a mother's blood during pregnancy affect the subsequent quality of attachment to infants. Mothers with a higher level of estradiol relative to progesterone during pregnancy had stronger feelings of attachment to their infants after the birth. Similar findings have been produced for other mammals. Of course, the conclusion that maternal attachment is influenced by hormones does not call for a rigidly deterministic interpretation. Even for simpler animals like rats, there is evidence that maternal behavior is affected by experience. If a female rat that has never been pregnant is exposed to rat pups, she initially shows little interest in them because she has not undergone the hormonal changes associated with pregnancy. Even so, her interest in the infants gradually increases. After about six days, she builds a nest, assembles the rat pups there and licks them. In fact, she demonstrates all of the behavior seen in a normal mother except for nursing the infants. Interestingly, male rats that normally do not care for the young behave rather like females in this artificial experimental situation, except that the females are more attentive foster parents.[20]

There is some indication that the stress of childbirth may also have some role to play in human mother–infant attachment. The evidence for this is that cortisol levels during childbirth are positively related to first-time mothers' attraction to infant odors. Cortisol is a stress hormone.[21]

Whatever the hormonal effects of pregnancy and childbirth, men and women have different levels of motivation for parental care. Even in childhood, girls are more interested in dolls, in play parenting, and in taking care of real infants. These differences emerge spontaneously and do not merely reflect sex-role training by parents. In fact, the most likely explanation is that sex differences in parental motivation evident in children are due to the early effect of sex hormones on the brain. Thus girls suffering from CAH syndrome, in which the brain receives abnormally high exposure to testosterone, have less interest in dolls and infants than other girls.[22]

Within two days after the birth of their child, mothers learn to recognize his or her cry. They respond to the cry of their own infant with a greater increase in heart rate than is produced by the cry of another infant. It is interesting that women who have not had a child show increased heart rate in response to the cry of an infant, something that is not seen in single men. The increase in heart rate is associated with elevated cortisol levels in the blood. Despite having male-differentiated brains and none of the hormonal priming of pregnancy and childbirth, men also develop responsiveness to the distress of their infants that is physiologically and behaviorally similar to that of the mothers. Many fathers have elevated cortisol levels and increased heart rates in response to the crying of their baby.[23]

The fact that men are capable of the same kind of responsiveness to infants as women might appear to call the role of hormones of pregnancy and childbirth into question, but this is probably a situation in which it is not one mechanism or the other, but both. Hormonal changes associated with pregnancy and childbirth may prime maternal responsiveness in the critical early days after birth. After some time, mere exposure to the infant tends to elicit parental responsiveness as a long-term guarantee that the infant will be cared for long after the effects of circulating hormones have subsided. For humans, as for rats and other mammals, parental care is an automatic response that is built up from several ingredients—developmental, hormonal, and experiential. Women are more strongly motivated to care for children because they generally receive a stronger dose of each of these components.

The sex difference in the work of childcare is thus arguably influenced by evolutionary adaptations, just as other differences in occupational preferences are. These differences are consistent with an ancestral division of labor between husbands and wives as they united their efforts to raise children, and they reflect adaptive sexual differentiation of the brain. Just because our brain biology has been shaped by the ancestral past does not mean that we are committed to doing the same thing in perpetuity, and time will likely bring many further changes in the occupational roles of women. One of the most remarkable facets of human evolution is the capacity to follow different strategies when confronted by different environments. Single parenthood is one example of an adaptive response to variations in the childhood environment and in marital opportunities.

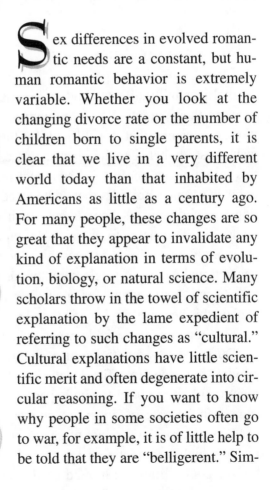

Chapter Nine

The Marriage Market and Single Parenthood

Sex differences in evolved romantic needs are a constant, but human romantic behavior is extremely variable. Whether you look at the changing divorce rate or the number of children born to single parents, it is clear that we live in a very different world today than that inhabited by Americans as little as a century ago. For many people, these changes are so great that they appear to invalidate any kind of explanation in terms of evolution, biology, or natural science. Many scholars throw in the towel of scientific explanation by the lame expedient of referring to such changes as "cultural." Cultural explanations have little scientific merit and often degenerate into circular reasoning. If you want to know why people in some societies often go to war, for example, it is of little help to be told that they are "belligerent." Similarly, if you ask why young women are more likely to have children out of wedlock today than fifty years ago, or why marriage has declined during the twentieth century, it is not illuminating to be told that they are "sexually liberated," or that there has been "a decline in moral values." These are basically liberal and conservative expressions of the same idea, unless we can provide an objective explanation for why such changes occur in a particular society at a particular time—that is what this chapter attempts to do by relating changes in the marriage market to single parenthood.

ROMANCE AND CHILDHOOD

Marital instability affects children because it alters the relationship between their parents. When the marriage market is unfavorable to women, meaning that it is difficult for them to marry men that they want, hostility between the sexes increases. Children thus learn from this to anticipate conflict in their own romantic lives. Their childhood, in essence, provides them with an education about marital relationships and prepares them to make their way in a similar emotional landscape. This helps to explain why children of divorced couples are more likely to divorce themselves and helps to account for cross-generational cycles of single parenthood. Many people would point to a genetic explanation for such differences, but genetics cannot explain this because single parenthood rates rise and fall steeply over time without any appreciable genetic changes in the population.

During the twentieth century, single parenthood rose to levels previously unseen in official statistics. Among African Americans, more children are born to single parents than to married ones. The same can be true of an entire country, such as Sweden. These changes are not the product of chaotic moral fads and fashions, as many social commentators imagine. Personal choices do matter, but the possibilities are limited by the marriage market and this is constrained by economic opportunities. Personal choices about sexual behavior are also affected by family background because the rearing environment gives children some powerful evidence about the realities of relations between the sexes.[1]

The quality of the relationship between children and their parents is of pivotal importance. Closeness to parents increases the probability of romantic happiness, as represented by a stable marriage and economic success. Lack of warmth in the parental relationship can produce irresponsible sexuality, in addition to other forms of problem behavior like delinquency, drug use, and low school grades.[2] Single mothers generally do not have satisfactory stable relationships with the fathers of their children. Children are sensitive to cues of parental conflict and to the lack of paternal support and commitment. This helps to explain why the children of single mothers are less likely to find romantic happiness in stable relationships when their time comes around. Their personal choices are affected by the romantic realities of their home lives. Young women are less likely to seek—and young men are less likely to provide—a deep emotional commitment to their partner before initiating sexual intercourse if this has been part of the pattern they observed in their parental generation.

The cross-generational cycle of single parenthood has been rationalized by evolutionary psychologists as reflecting an adaptive mechanism that helped our hunter-gatherer ancestors to adjust to the different social conditions they encountered. As already noted, children raised during times of famine, for example, would have had to be much tougher and more competitive to survive than children raised in more plentiful times. Perhaps the clearest modern example of this adaptive mechanism in operation concerns the differing outlook and trust among children raised in low-social-status homes compared to those raised in comparative affluence.

Children raised in poorer homes are much more likely to be the recipients of verbal hostility from caregivers, and they are much more likely to respond in kind. Young women raised in such socially trying environments are likely to suffer from a perceived lack of emotional closeness with their families that provides an incentive for early sexual activity and makes the raising of a child, who can be loved, seem an attractive prospect. Young men raised in coercive homes are strongly motivated to exploit young women to gratify their sexual needs.

SINGLE PARENTHOOD AND MARRIAGE PROSPECTS OF AFRICAN AMERICANS

Despite white single parenthood rates that rose steadily during much of the second half of the twentieth century, African American teenagers are more likely to become single mothers than the rest of the U.S. teen population. According to 1990 census data, the annual nonmarital birth rates of blacks aged fifteen to nineteen years was 10.7 percent compared to 2.7 percent for whites in the 129 metropolitan areas with black populations greater than twenty thousand. These areas were analyzed by the author in a study of predictors of group differences in single parenthood. (Note that since there are five years in the age group, these annual rates are multiplied by five to estimate the proportion of teenagers giving birth, which was 54 percent of blacks and 14 percent of whites.) The annual rates can be compared with teen birth rates of 5 percent in the United States as a whole, 2 percent in western Europe, and 6 percent, on average, for all countries in the world. Such comparisons present a technical problem because there is a distinction between teenage childbearing and single parenthood. Yet this technicality is less important than might be imagined for two reasons. First, most teen births are to single mothers in this country. Second, even when teens marry, their marriages are more unstable than marriages contracted by more mature couples in their twenties. From a practical perspective, teenage childbearing thus raises similar problems for children whether the teen mother is married or not.

The most important questions to ask about group differences in single parenthood are (a) whether these can be explained in terms of differences in the social environment, particularly the marriage market, and (b) whether the causes of group differences in single parenthood are also responsible for changes within a group over time.

Contrary to the views of moralists, single parenthood is not a voluntary choice so much as a response to weak marriage market conditions in which marriageable men are scarce. Survey data have shown that young black women want to marry just as much as their white counterparts, but they do not have the same marital opportunities.[3] For practical purposes, the universe of marital possibilities for black women is circumscribed by the availability of marriageable black men. If they marry, it is almost always (i.e., 97 percent of the time) to black men.[4] Marriage within their ethnic group is slightly higher for African American women, and this might reflect a universal tendency for brides and grooms to resemble each other in a whole range of attributes such as background and interests. Racial concentration of neighborhoods might also be important because it is easier for people to get to know each other and form romantic attachments if they live close together. Chance encounters are the stuff of romance and such meetings require physical proximity. This would help to explain why people residing in large, supposedly anonymous cities are more likely to marry if they live within a few blocks of each other.[5] For a variety of reasons, a black woman's marital prospects are strongly determined by the availability of marriageable black men in her neighborhood.

These limitations on marital opportunity have exposed contemporary black women to a severe marriage squeeze. The scale of this problem was clear by 1980. According to census data for that year, for women aged twenty to fourty-four years (the prime marital ages), there were only 86 African American men for every 100 women. This means that approximately one-sixth of young black men had apparently died from accidents, illnesses, murder, suicide, or the Vietnam War, given that there are slightly more males than females at birth (about 3 percent more).

A high early mortality rate for black men is far from being the whole story in respect to the marriage squeeze. Of the survivors, many were poor marriage prospects for various reasons. Around 2 percent were locked up in prisons, which made them practically inaccessible, leaving 84 men for every 100 women. Of these available men, 30 percent did not participate in the labor force, further shrinking the pool of marriageable men to 54 for every 100 women. Subtracting the 8 percent who worked but were currently unemployed left only 46 marriageable black men (i.e., those in steady employment) for every 100 women.[6] Even this number might have been inflated because it included homosexual men, and there are always more male than female homosexuals.

Grim as they are, the numbers in themselves do not convey the true dimensions of the problem. Markets, including marriage markets, are dynamic. When there is such an imbalance between the demand and supply of marriageable black men, the marriage market is not merely difficult but tends to disintegrate entirely. We see this in the phenomenon of many young African American women involuntarily settling for early reproduction outside marriage. The window of opportunity for marriage is also compressed in time with younger, more physically attractive women having an advantage. With marriage prospects so bleak, many older women must become single parents if they are to have children at all.

Parental Investment Prospects

Some moral critics condemn single mothers for raising children in poverty and increasing the probability that economic difficulty and its attendant social problems will be passed on to the next generation. Needless to say, these fears have some basis in fact. Children of single teen mothers are more likely to be poor as adults, more likely to be single parents, and more likely to be arrested for criminal activity.[7] Given that single parenthood is not usually voluntarily chosen but is a response to diminished romantic opportunities, moral condemnation misses the larger picture, however. Scarcity of marriageable men who are able and willing to help in raising children forces women to reproduce outside marriage.

Yet the marriage market is not the entire story. Given the same marriage market, not all black women opt for early reproduction outside marriage, and moralists would applaud this. Yet even this narrow opportunity for moralistic interpretation quickly closes when we obtain a grasp of the psychological factors at work and their roots in childhood.

From a psychological perspective, single parenthood can be a reaction to lack of emotional closeness among family members during childhood. Investment by fathers is often diminished by virtue of the father's physical absence or emotional distance from children. Young inexperienced mothers may feel a great deal of emotional strain in caring for their children and this produces anger and hostility directed at children. According to this perspective, the sexual relationships of teen mothers and their attraction to early motherhood are each based on the same desire for emotional closeness with other people. An emotionally stressful rearing environment makes young women more likely to be sexually active from an early age and to want children. A high need for love goes along with low career aspirations, as though the emotional investment that young women make in an occupation subtracts from their interest in starting a family.

The most direct evidence supporting such a connection is the fact that poor academic performance in high school is one of the most powerful predictors of teen motherhood, with low achievers being five times as likely to become pregnant as high achievers. Young women with high career aspirations are less likely to be sexually active and if they are having sex, they are more likely to use contraception. In the event of an unplanned pregnancy, they are more likely to have an abortion than is true of teens having low aspirations and/or low school grades.[8]

It is often pointed out that adolescent pregnancies occur more by accident than design. Yet the unwillingness of young women with poor career prospects to have an abortion is one of the clearest indicators of their pronatal intentions. Careers and babies are hard to reconcile even for older, more mature women; this is most clearly demonstrated in the fact that career women postpone reproduction for a decade compared to those working in subsistence-wage jobs.[9]

The dampening effect of career motivation on sexuality and reproduction is not just a social class difference. Even among affluent young women, those who are interested in casual sex have poor grades. Their agreement with the statement, "I could be comfortable having sex with different partners," provided far better prediction of their college grades than scores on the American College Test (ACT, like the SAT) which had been designed by testing experts for this purpose.[10] This is consistent with the finding that teen mothers, in general, have poor high school grades.

Thus, it might seem that career ambition counteracts the desire to have children, whether this is due to the psychology of need satisfaction, or whether it merely reflects the practical difficulty of establishing a career while taking care of children. An alternative point of view is that the backgrounds of teen mothers simultaneously undermine academic achievement and motivate teens to be sexually active and give birth. Thus, poverty is associated with low academic attainment and high teen birthrates.

Being raised in poverty undermines academic performance for many reasons that are now quite well understood: there is often little in the way of early brain stimulation because of the absence of enriching toys and books, and also because these children are not read to and have little in the way of positive verbal interaction. Attendance at second-rate schools goes along with growing up in a poor neighborhood because schools are run based largely on local tax support and increase in quality with the income of a neighborhood. Perhaps just as important is that children raised in poverty may not acquire much respect for academic pursuits, either from their parents or from their neighbors. Although education may be valued in the abstract, parents typically do not or cannot make much effort to support their child's learning—for example, by arranging quiet times and places for homework to be done. In the jargon of social science, poor homes are often not "educationally efficient."

Poverty is psychologically stressful. This is certainly true of the lives of children who are exposed to a great deal of interpersonal aggressiveness and experience a lack of control over unpleasant events in their lives—having the electricity turned off, being evicted from their homes, or encountering drug dealers and other dangerous criminals each day. Study of the home lives of teen mothers indicates that they are exposed to many stressors. Examples include: lack of male emotional or economic support; parental conflict; divorce or emotional distance from fathers; physical, emotional, or sexual abuse; living in foster homes; and drug addiction or alcoholism of one or both parents. Lax parental supervision after school is also an issue because this provides an opportunity for sexual behavior in the teens' homes. The loss of contact with their fathers during childhood—more likely for poor children—gives children a sense of emotional deprivation that makes early sexual relationships more appealing.[11]

Exposure to family violence is a real problem in the homes of a majority of adolescent mothers;[12] evolutionary psychologists see this as significant because it tells young people that the social environment is harsh and unstable. Under such conditions, they cannot expect to receive much support from their mates. Delaying reproduction until marriage to a desirable spouse would thus have made no sense. Our female ancestors would have increased their reproductive success under such conditions of low paternal investment if they began reproducing early in life, thus capitalizing on youthful vigor and health.

Teen mothers are likely to have witnessed violence against their own mothers or to have been victims of violence themselves. They are also vulnerable to sexual abuse. A study of 445 pregnant teens and teen mothers in the Illinois-based Ounce of Prevention program found that 61 percent had experienced sexual abuse, compared with a national incidence of around one in four. Vulnerability to sexual abuse, particularly by other family members including brothers, grandfathers, uncles, and mothers' boyfriends is increased by the absence of fathers from the home.[13]

In addition to receiving little emotional support from their fathers, teen mothers often have a difficult relationship with their mothers, who may be emotionally distant and rejecting or emotionally dependent on their daughters. This problem is sometimes worsened by the lack of a social support network outside the family.

Poverty is not romantic and it might seem that growing up in an abusive household with a great deal of conflict between the sexes would make love and marriage unappealing. In some ways it does, but such is the mysterious nature of life on earth that love conquers rationality. Witnesses of marital conflict may be more, not less, drawn to the opposite sex.

Teen mothers generally have conflicting attitudes toward men, according to interview research. Most see men as potentially violent and alcoholic. Despite this, they often entertain excessively romantic expectations of their own dating partners. Many become pregnant deliberately in the fond belief that their boyfriends will fall hopelessly in love with them and desire marriage. The probability of such marriages occurring is low and the prospects of these lasting are even lower. A commonly expressed motive for becoming pregnant and giving birth is to have someone to love. Young women with a strong desire to produce a baby are less likely to use contraception and less likely to seek an abortion in the event of becoming pregnant.[14]

Inconsistent contraceptive use is clearly a risk factor for teen pregnancy. Another factor is the early age of engaging in sexual intercourse. Age of first consensual intercourse is itself predicted by early age of sexual maturation as well as having been a victim of sexual abuse. Absent fathers make young women more vulnerable to sexual abuse and might even advance the age of sexual maturation by a few months at least.[15]

Most of the identifiable risk factors for teen pregnancy are higher for African American women. This applies to onset of sexual relations. A 1987 study found that four times as many fifteen- to sixteen-year-old African Americans were sexually active, compared to whites of the same age.[16] More recent research has confirmed this finding.[17] Another interesting group difference is that black teen mothers often lost their virginity in a casual relationship. By contrast, Mexican American teen mothers often have had sex only with the father of their child, with whom they have had a long-term relationship.[18] This difference probably arises from the better marriage market for young Mexican American women. Precocious sexuality is a feature of societies in which many young people do not marry, such as Sweden (see chapter 10).

Even though we are all individuals and our lives are shaped by unique events and idiosyncratic motives, there are a few themes that recur in studies of teen motherhood in our society. Many teen mothers suffer from a lack of emotional support from other people in their families. They struggle in school and do not want to pursue careers requiring many onerous years of education. Their career motivation is low, and their need for love in sexual relationships and childbearing is high. For African American women, the bleakness of the marriage market offers little chance that they can satisfy these needs through marriage, or even have a long-term boyfriend living with them. For them, motherhood outside wedlock has become the statistical norm, even though it is not seen as desirable. Placed in the cruel bind of either not raising children at all, or raising them out of wedlock, most choose the second option.

Single teen parenthood is predictable from the practical choices faced by young women; this conclusion comes from statistical tests conducted by the author to account for differing teen birthrates in 185 countries around the world. Adolescents in various countries are most likely to become teen mothers if there is a scarcity of men of an appropriate age for marriage. Economic conditions are also important because these determine the job prospects of men. Teens are more likely to give birth in poor countries where there are fewer men with steady jobs commanding a large enough income to support a family. Early reproduction is a response to poor marriage market conditions because if there is little chance of contracting an economically beneficial marriage, there is no good reason for delaying reproduction, as it is unlikely that the marriage market will improve quickly enough to make a difference.[19]

Given that *national* differences in teen birth rates can be statistically explained in terms of marriage market conditions, is it possible to explain *ethnic* group differences in the United States in similar objective terms? The group differences are large, but study of their history indicates that there is no intrinsic reason why African Americans should have higher teen birthrates. Thus, in 1993, the illegitimate birthrate (per 1,000 single women) was 36 for whites, exactly the same as the rate had been for blacks in 1940. The white percentage has subsequently increased. The author recently examined black-white differences in nonmarital births from 1955 to 1994.[20] Could group differences in single parenthood for teens be explained in terms of parental investment prospects, as defined by the availability of men of marriageable age and economic conditions? Could changes in the marriage market explain the strongly increasing trend of illegitimacy for both blacks and whites (statistical categories defined by government statisticians) during the second half of the twentieth century?

The proportion of men to women aged twenty-five to forty-four years was used as a measure of the marriage market for both groups. The current marital status of members of each population was not considered because previous research had shown that even though people are currently married, they still act as "players" in the marriage market and are capable of divorcing and remarrying if they perceive an improved marriage opportunity. This is particularly true when divorce rates are high, as they were during the time of the study.[21]

My study also included women's employment rates as an indirect measure of male economic support for the marriage. The logic is simple: if men are unable to provide for their families, then women go to work. This logic is also controversial because women go to work and develop careers for individualistic reasons, such as the desire to fulfill career goals and improve their living conditions. Yet historical evidence suggests that the marriage market has a powerful influence on women's career aspirations. The most compelling indication involves the post–World War II period in the United States. Women who had entered many masculine occupations such as shipbuilding during the war quickly abandoned them in favor of marriage when a large number of eligible single men returned from fighting on foreign battlefields. During the prosperous decade of the fifties, women stayed away from careers in droves because blue-collar wages were high enough to support a family in unprecedented luxury. As the marriage market worsened, and as economic conditions deteriorated in the 1970s, more and more women entered the paid workforce and the trend continued up to the present. During the same period, women have become more career-oriented and are more often raised to pursue career goals than was true in the past.

Women who are apart from their children for most of the work day clearly cannot invest as much time in their own children and must rely on commercial surrogates (i.e., daycare workers) to do the job. Although the consequences of daycare are controversial and presumably vary greatly with quality of the facility, women's work participation may reduce their direct investment in children. Reduced parental investment—as a general rule, not in every case—increases social problems such as drug abuse, teen pregnancy, and criminal involvement. It is probably no accident that the historical period of rapid increases in American female labor participation has seen huge increases in all of these problems. If male economic investment in families is reduced, the number of children living in poverty rises. For this reason, the proportion of families living in poverty was included in the study as a measure of the unavailability of men who are economically viable marriage prospects. It was predicted that single parenthood would increase in response to a smaller pool of marriageable men and their diminished capacity to invest in children (as reflected in high poverty rates and high female labor participation).

Analysis of the government data showed that "illegitimate" births increased as the number of men of marriageable age declined, as poverty rose, and as women's labor participation increased. When these variables were entered in a regression analysis along with the year (used as a control variable), they provided almost perfect prediction of single parenthood, explaining 97 percent of the variation in teen pregnancy rates over time and across racial groups. The analysis also found that ethnic group differences in teen births could be entirely explained by measures of parental investment, i.e., by the social environment. This clearly contradicts the notion advocated by Canadian social psychologist Philippe Rushton, among others, that group differences in social problems, like single parenthood, are based on intrinsic biological differences.

Yet, it might be a mistake to read too much into correlations over time, which could be accidental. The findings were thus tested using a different research strategy. Teen birthrates for blacks and whites in U.S. states were analyzed for the same year.[22] Parental investment helped account for differences in teen birthrates between states. It also completely explained black–white differences in teen birthrates. These findings were replicated using 1990 census data for U.S. metropolitan areas having African American populations above twenty thousand.

When the African American data were considered separately in the metropolitan areas study (which did not include the more affluent suburbs), differences in the marriage market for blacks in various cities no longer accounted for teen birthrates. Evidently, so few young women are marrying in these communities that the marriage market is in trouble. The overall percentage of young women who became single mothers was 54 percent.

The decline of marriage has two components. One is obviously the scarcity of marriageable men, as already mentioned. The other is the fact that single men have a reduced interest in marriage. (It should be emphasized that these results do not apply to distant suburbs to which many upwardly mobile African Americans have moved to escape the distressing environment of the inner cities, however.)

RELATIONS BETWEEN THE SEXES

A soft marriage market for women produces hostility between the sexes and creates a sense that men and women are at cross-purposes. This is particularly true of black men and women of the inner cities.

Social scientists often attribute this to generalized social disintegration. Elijah Anderson, an urban anthropologist, focuses on relations between men and women, and highlights the absence of male commitment to long-term romantic relationships or child support among some of those in the inner city.[23] He sees dating reduced to a cruel game of emotional manipulation; in this sexually divided society, some young women are impregnated and raise children by themselves while some young fathers hang out on street corners bragging about their most recent sexual exploits.

Anderson designates this group of young sexual predators as "streetwise." This means being ruthlessly Machiavellian in securing one's goals regardless of the feelings of or injury to others. The hard grind of a job or conventional career is rejected by this group in favor of criminal enterprises where easy money is made from drug trafficking, car theft, prostitution, mugging, and illegal gambling rackets. Streetwise men pride themselves on a capacity for seducing women by exaggerated yet insincere pledges of affection, love, and commitment. Being streetwise means that caring for one's children is not even a remote possibility. As soon as a woman becomes pregnant, she is routinely abandoned and the father denies that the baby is his.

As many social commentators have remarked, there has been a real decline in two-parent African American families in recent history. Thus, between 1960 and 1990, the rate of out-of-wedlock births increased by a factor of three, mirroring the increase in the

U.S. population as a whole, although beginning from a much higher value. Even though teen birthrates were much lower in 1990 than in 1960 (9.6 percent each year versus 15.5 percent), reflecting a general decline in the birthrate, the rate of teen births to single mothers was over three times higher.[24] Teen birthrates for African Americans are still high at 10 percent compared to 6 percent, on average, for the all countries in the world, and 1 percent for western Europe.[25]

Fewer teens marry today when they become pregnant. One reason is a decline in the dubious institution of the "shotgun" marriage, where young men who impregnate teens are coerced or actually compelled to marry them. In countries with a strict code of honor that revolves around premarital chastity, like Pakistan and Jordan, this is done on pain of death. Among African Americans, the decline of the shotgun marriage is connected to the marginalization of men. Elijah Anderson observes that this group of streetwise men can exploit young women at will because the young women are unprotected by male relatives.[26] Black teenage women cannot rely on the support of fathers because fathers are unlikely to be living in the same home. Many conservative scholars see the decline of the black family as a symptom of a deeper social pathology, but this moralistic argument explains very little. You are still left with the problem of explaining why social pathology emerges at some times and in some places but not in others. The real reason has to do with changes in the marriage market.[27] When it is difficult for women to marry, they are also vulnerable to being sexually exploited.

SEDUCTION GAMES

Interestingly, most young black women would like to marry and do not see single parenthood as desirable. Teen mothers do not always intend to become pregnant outside wedlock. Sexual favors, as in many societies, are sometimes used to manipulate men. Sex is a bargaining chip used to compete for the attention and love of a desirable partner. The teens often failed to use contraception because of the vain belief that they would not become pregnant—but if they did, they would have someone to love. They also wanted to believe that if they became pregnant, their lover would live up to his promises of love and marriage.[28]

The causes of high rates of single parenthood among poor African American teenagers are no longer a mystery. There has been a steep decline in marriage among teenagers so that, by 1996, 95 percent of black teen mothers were unmarried (compared to only 15 percent in 1960).[29] The weakness of the marriage market is due to a scarcity of black men of an appropriate age who can contribute dependable economic support to a household. The marriage market is so slanted against women that for many there is little prospect of matrimony. Because there is little point in waiting for Mr. Right, some young and indigent women begin their reproductive careers early. These objective realities shape the romantic interactions and sexual behavior of young black men and women, and are responsible for the distressing pattern of conflict between the sexes described by Elijah Anderson. At their core, these problems are economic, and improving economic conditions would reduce the problem of single parenthood by elevating more bachelors to marriageability. Economics affects many other aspects of relations between the sexes, including perceptions of the female body build that is considered attractive.

Chapter Ten

Fashion Trends and Marriage

Anorexia, Beards, and Skirt Length

Some aspects of physical attractiveness phenomena are etched in the brain before birth, such as the preference for beautiful faces or attraction to the body shape of the opposite sex. Others vary widely from one society to another and even within the same society over time. We are as much enslaved to the standards of attractiveness we *acquire* from being raised in a particular society at a particular point in its history as we are to the *evolved* ones with which we are born. Responding in this way to the prevailing fashions helps us to achieve social and economic success in different social environments because fashions of physical appearance change in ways that make adaptive sense. For example, plump people are admired more in societies where food is scarce than where food is plentiful: in such societies plumpness is a sign of social success, but the opposite is true of our society, where food is generally available. Evolved responses to fashion standards thus help us to be successful in careers and marriage in different social environments.

Fashions can run to extremes. This is certainly true of the slender standard of bodily attractiveness for women that has taken hold in industrialized societies as women compete for entry to professions. In these societies, women want to be slim to create an impression of professional competence. However, the need to be slender can produce fatal eating disorders.

A very public example of this phenomenon is Crown Princess Victoria, heir to the Swedish throne.[1] A hefty teenager, Victoria was active in sports but experienced strong pressure to set an example to others by losing weight. The popular press was fond of criticizing her for looking as though she had consumed too much fast food. As though in response to her critics, Victoria began to lose weight and before long was featured in a bikini on the cover of *Svensk Dam Tidning*, a women's magazine that had found fault with her appearance in the past. The weight loss was too great to be healthy and her appearance in an evening gown on the occasion of her twentieth birthday showed the public that she had become too skinny. Within a week, the palace confirmed that she was suffering from anorexia nervosa.[2]

Anorexia is not a very common problem, striking about 1 percent of young Swedish and American women (although other eating disorders such as bulimia are much more common). Yet the fact that living up to the prevailing standard of attractiveness can be confused with a fatal illness shows just how extreme the slender standard is in economically developed countries. How could this have happened?

FAT AND FERTILITY

Contrary to the slender standard prevailing in our society, the anthropological literature is full of examples of an appreciation of female fatness. Young American women find these attitudes gross and incomprehensible. How could they agree with the claim of the Siriono tribe inhabiting the rain forests of Brazil that an attractive woman must be corpulent, with huge breasts and broad hips? Equally incomprehensible are the Enga, natives of New Guinea, who feel sorry for slender girls because they have little chance of marrying the man of their dreams. Enga girls participate in a simple morning prayer ritual designed to ensure that they will grow sleek and fat. Among the Enga, oily skin is one of the most important criteria of physical attractiveness in woman. Fatness is attractive for young women in some societies because people associate it with fertility and, interestingly, the work of endocrine physiologists supports this view. Women who are very physically active, like professional dancers, lose much body fat and sometimes experience the loss of menstruation and infertility. Their bodies respond to diminished fat stores as though they were experiencing famine: reproduction is temporarily shut off in the interests of saving energy and promoting survival.[3]

It is not hard to understand why people in subsistence economies would find thinness unattractive in young women. They associate slenderness with difficulty in conceiving. Moreover, in the past, when breast-feeding went on for several years, maternal fat stores played an important role in the survival of children because these stores were drawn on during food shortages. If mothers were slender, they would have had trouble producing enough milk to raise their children to nutritional independence. This is one reason why, in societies where hunger is a daily reality, people are more likely to prize female fatness. Of course, the unhealthy levels of extreme overweight found in modern affluent countries were never a feature of subsistence economies: these societies had greater difficulty in obtaining large quantities of high-calorie foods and, possibly, greater physical activity associated with a hunter-gatherer lifestyle.

The attractiveness of body fat for young women is illustrated by the custom of fattening the bride in many preindustrial societies. Young women of Sierra Leone gorge themselves on vegetable oils and other fattening foods. Tamil brides, in Sri Lanka, are fattened with meals of raw eggs, steamed rice, eggshells, vegetable curries containing sesame oil, grated coconuts, sweets, and cakes. They wash all of this down with drinks of margosa oil.[4]

Confronted with such extremes in different societies, it is no wonder that students of fashion should conclude that it is intrinsically arbitrary and unpredictable. Yet they are wrong. Fashions of bodily attractiveness for women vary predictably with female economic and political power.

THE SLENDER STANDARD IN HISTORICAL CONTEXT

Women's desire to be slender might appear to be a uniquely modern phenomenon. Yet there is fairly compelling evidence of a worldwide pattern that helps us to understand why some societies produce eating disorders and others admire a heavier standard of feminine bodily attractiveness.

This pattern was reported by Judith Anderson, an evolutionary psychologist at Simon Fraser University in Canada, in the context of a study based on preindustrial societies described in the Human Relations Area Files. Anderson found that slenderness is admired in societies where women are socially powerful, as measured by the perceived importance of their work, property ownership, and political influence.[5] Women are less powerful in societies that admire plumpness.

Fat that is stored in the body can be thought of as a store of energy that can be drawn upon in times of food scarcity. This would explain why it is valued in subsistence societies and accounts for

Fig. 11. The Mother, a painting by Gari Melchers, captures the altruistic, or self-sacrificial, aspect of breast-feeding. (Reproduced from the Collections of the Library of Congress)

the practice of fattening the bride. Plump women are more likely to conceive, and they can support the great energetic demands of pregnancy and breast-feeding (see fig. 11) that goes on for at least three years in preindustrial societies.[6]

While we may be surprised at the range of variation in body fat that is considered attractive for young women in different societies around the world, there has also been an astonishing amount of variation in what has been considered attractive at different times in America within the past century. Whether women want to be slim, voluptuous, or something in-between is apparently determined by a compromise between careers and marriage as economic strategies. During historical periods when women enter careers in large numbers, as happened twice in the twentieth century, the standard becomes more slender. At the same periods, the importance of marriage declines.[7]

The standard of bodily attractiveness for women changes markedly from one decade to the next, as can be seen by studying the bodily dimensions of models portrayed in ladies' magazines such as *Vogue* and *Ladies' Home Journal*. While these publications have a different female readership, with *Vogue* catering to a more affluent market segment, the bodily proportions of models depicted in each agree almost exactly from year to year. Changes in the standard thus occur simultaneously for middle-class and upper-class readers. This nearly perfect agreement over time suggests that the publications are representative of the standard of attractiveness held by most American women, rather than just reflecting the arbitrary subjective decisions of editors. In 1900, models shown in both magazines had bust dimensions that were twice as big as their waists. This means that a model with a tiny twenty-inch waist would have had a forty-inch bust (see figs. 12 and 13). This unnatural appearance was maintained using tight corsets. During the early 1920s there was a pronounced shift toward slenderness, and by 1925 the bust was only 1.1 times the waist dimension. Flapper women were admired for having figures that did not differ greatly from the typical man (see fig. 14). This ideal standard was even more slender than that of the 1990s, when popular actresses like Julia Roberts had bust dimensions that were about 1.3 times the size of their waists (e.g., a thirty-three-inch bust and a twenty-five-inch waist).[8]

Everyone knows that while the skinniness of contemporary

Fig. 12. Lottie Collins, an entertainer, photographed in 1892, illustrates the narrow waist and ample proportions of attractive women in this period. (Reproduced from the Collections of the Library of Congress, LC-USZ62-101298)

Fig. 13. These curvaceous ladies were typical of fashion models in 1906. (Reproduced from the Collections of the Library of Congress, LC-USZ62-095360)

Fig. 14. This evening gown, from 1921, de-emphasizes the feminine body shape (see text for explanation). (Reproduced from the Collections of the Library of Congress, LC-USZ62-100835)

fashion models may be natural for a few women, it is a dangerous, or unattainable extreme for the majority. The slender standard has motivated women to starve themselves and engage in demanding exercise programs without adequate nutrition. In the 1920s it produced an epidemic of eating disorders that ended quickly with the Great Depression when there was less food available. Shrinking job opportunities for women were accompanied by the return of a voluptuous standard of bodily attractiveness. Once again, women were forced to seek the economic support of husbands. The current slender standard is much more persistent, having continued from the 1960s, and it reflects consistently good job opportunities for women (as well as diminished marriage prospects). As a result, there has been a long-lasting epidemic of anorexia and bulimia primarily afflicting young single women. In the past, eating disorders were more common in women from affluent homes, who would have been more career-oriented. Today eating disorders strike women of all social classes, presumably because of the increased importance of occupational success for women of all origins.[9]

EXPLAINING THE SLENDER STANDARD

Why are eating disorders tied to the job market? Working in the mid-1980s, sociologist Brett Silverstein[10] noted that highly curvaceous women are unfairly perceived as lacking professional competence because of a stereotype that sexually attractive women cannot be intelligent.

There is little doubt that this stereotype exists. In one experiment published in 1980, women wore padded bras to simulate various levels of curvaceousness and their pictures were taken for use as stimuli in the research. Students evaluated the women who seemed to be large-breasted as lacking in professional competence and inferred that they lacked sexual restraint. The experiment confirmed that there is a "dumb blonde" stereotype of a sexually attractive woman as lacking in competence and intelligence.[11] The same stereotype is alive and well today, and recent research has demonstrated that having a curvaceous body has much more negative implications as far as evaluating a woman's competence is concerned than does being moderately overweight.[12]

When women are entering business and the professions in large numbers, they need to seem more professionally competent, which means understating their stereotypically feminine traits. At such times, the standard of bodily attractiveness should become more slender and less curvaceous. Brett Silverstein found support for this hypothesis in his study of the dimensions of models depicted in ladies' magazines. As more women entered careers, the standard of bodily attractiveness became less curvaceous.[13] This explains why continual gains by women in professional status and earning power over the past four decades has gone along with a persistently slender standard of attractiveness and eating disorders, including anorexia and bulimia, affecting at least 10 percent of American women.

Instead of being an arbitrary phenomenon, evidence has accumulated that varying standards of feminine bodily attractiveness are associated with many measures of economic growth, and with marriage. The standard becomes increasingly slender as the economy improves, as indexed by stock market indexes, and income per person. When the economy expands, women's job opportunities improve, creating pressure for them to seem professionally competent. A strong economy thus means a slender standard.[14]

In spite of the variable nature of fashions of attractiveness, both male and female college students agree that curvaceous women are sexy-looking and that they would be good at reproducing. Medical evidence corroborates these intuitions. Curvaceous women become pregnant more easily. (This is not speaking of obese women.) Curvaceous women have fewer complications during pregnancy and delivery. Their general health is also much better, with a reduced vulnerability to most serious disorders, such as cancer, heart disease, and mental illness. As far as heart disease is concerned, this is associated with being overweight, but people are at greater risk if fat is stored around the midriff, as opposed to curvaceous women who store fat at other sites. These health advantages are attributable to curvaceous women having a high level of estrogen relative to testosterone circulating in their bodies. This surprising conclusion has a simple explanation: the high estrogen levels that sculpt the feminine body into its hourglass shape by controlling the deposition of fat also boost the immune system, resulting in better health.[15]

In the evolutionary past, men who selected brides with highly curvaceous bodies would have enjoyed a reproductive advantage. Their wives would have conceived more quickly and would have had better health to raise their offspring to maturity. This produced a selection pressure for men to be attracted to women with stereotypically feminine body shapes. The role of evolution in the esthetics of feminine beauty is suggested by the fact that what men are attracted to in the female body changes surprisingly little given the dramatic swings in fashion as depicted in women's magazines. This conservative tendency (i.e., lack of change) is illustrated by the fact that models depicted in *Playboy* and pornographic magazines never approach the extreme skinniness and angular form typical of fashion magazines—although English men of the Victorian period in England admired heavier women than English men do today, for example.[16] Currently, women want to be more slender than men want them to be, according to various experiments conducted by psychologists. Women strive to be slender because this affects their ability to seem professionally competent and compete for good jobs.

There is a difference between what women consider the ideal standard and the reality of their own bodies; this disparity is responsible for a great deal of dissatisfaction on the part of women who cannot be as slim as they would like to be. Despite the fashionable slender standard of today compared to the heavier ideal of the past, when women were less career-minded, modern women are actually heavier than they used to be. This mismatch between the ideal and the reality likely reflects two trends: increased height due to improved nutrition during childhood, and increased storage of body fat. Obesity is widely recognized as a serious health problem in the United States, and it can be attributed to the combined effects of consuming high-calorie items, such as snack foods and fast foods, and a sedentary lifestyle. The gap between the ideal and reality causes misery and eating disorders at all levels of society, from the poor to celebrities.

One interesting discovery of psychologists is that the perception of women's body build by men and women (as distinct from their ideals) is closely similar. Both agree that curvaceous figures are sexy and that slender ones project an image of competence. Moreover, curvaceous women are seen as incompetent. If a woman's figure conveys a message of sexual attractiveness, it cannot be simultaneously seen as competent, according to the research. Conversely, if a woman's figure creates an impression of professionalism, it cannot simultaneously be perceived as sexy. This means that there is always a compromise between sending one message and sending the other. The basis of this compromise is very familiar to women who must seem professional in their self-presentation. Women who happen to have voluptuous figures, for example, can play down this message by wearing tailored business suits that make them seem more angular. A woman who adopts "corporate" dress style at work may wear far more seductive clothing when dating.[17]

Another implication of the compromise between appearing competent and appearing sexy is that the ideal figure for women at any point in history is affected both by the economy and by the marriage market. When women enter the professions in droves, as happened in the 1920s and from the 1960s on—they want to seem competent, and the ideal figure becomes slender. During the 1920s, the standard became so slender that there was a serious outbreak of eating disorders, although it was brief, ending when the bottom fell out of the job market following the 1929 stock market crash. In the subsequent Great Depression, women found it very difficult to enter careers. The diet-and-exercise craze of the 1920s was similar to that of today except that the many advertisements for "reducing girdles," potions, and machines seem archaic today. Yet the 1920s produced exercise films that featured well-known actresses of the period.[18]

If marriage prospects are unusually good (i.e., many more young single men than young single women) and women choose marriage over careers, as happened in the 1950s, the standard of attractiveness moves back to the fuller figures popular in the 1950s, such as that of actress Jayne Mansfield. When women rely on husbands for much of their economic support, the ideal body for women is much the same as that preferred by men. The ideal body depicted in fashion magazines thus has little relationship to the actual body build of most real women. Instead, it can be thought of as a preferred build at which women are most successful when competing economically with other women. If most of their income comes from their occupations, the ideal is very slender. If most is derived from husbands, the standard is much more curvaceous. If women rely on a mix of occupations and marriage, as happened in the 1950s, the standard of attractiveness falls somewhere between these extremes. It is interesting that we see popular actresses of this period, like Jayne Mansfield, as very curvaceous but they were not so extreme in this respect as female entertainers fifty years earlier.[19]

For the past several decades, the strong economy has allowed women to focus on careers rather than marriage so that the standard of attractiveness has remained slender. The same period has seen the emergence of dress styles that emphasize autonomy and sexual liberation of women.

Dress Styles

Since Western women cover a good part of their bodies with clothing most of the time, it seems obvious that changes in standards of bodily attractiveness would have an impact on dress fashion. Consistent with this expectation, changes in women's dress styles have also followed the trends of marriage and economics. This provides further evidence of the connection between fashion and economic competition between women. Dress styles can amplify or minimize the curvaceousness of women's bodies. Moreover, they transmit signals about behavioral intentions. For example, the extensive covering of the body, and even the face, by women in traditional Moslem societies sends a message of sexual inaccessibility that is amply backed up by harsh social strictures against premarital intercourse. Similarly, the miniskirt that became popular in the 1960s transmitted a message about sexual freedom and accessibility. Why women at different times and places should convey such radically different messages about their sexual intentions has always been somewhat of a mystery, but it has begun to make sense when considered in terms of their economic strategies, i.e., the relative importance of jobs versus marriage as forms of economic support.

When women are pursuing marriage as a primary strategy, they emphasize health and capacity to reproduce by adhering to a curvaceous ideal. At the same time, they paradoxically advertise sexual modesty in their dress. It is as though they were saying to prospective husbands, "I am extremely desirable but difficult to obtain!" Sexual modesty is of considerable importance to men in selecting a spouse in many societies because it provides a behavioral guarantee of sexual faithfulness. For most men, the prospect of sexual infidelity in a spouse is extremely undesirable because of an evolutionary psychology of sexual jealousy that protects them against being cuckolded and raising other men's children.[20]

The messages transmitted by clothing style are, to some extent, arbitrary. To take the most obvious case, people in subsistence economies inhabiting warm climates may wear little or no clothing except on ceremonial occasions. For some reason, they obtain no advantage from modifying natural bodily signals related to physical attractiveness. Perhaps the labor of garment-making outweighs any of the communicative advantages of dress, whether clothing is used to exaggerate physical attractiveness or to conceal physical defects. It is rather interesting that no technologically advanced society has ever gone from wearing clothes back to nakedness even when the weather is warm enough for this. Even the nudist beaches that provide a limited exception have failed to catch on in most developed countries. Clearly there is a social advantage of controlling bodily signaling through dress. Clothing not only alters evolved bodily signals but transmits messages about occupation and social status in complex modern societies.

An adaptationist approach to fashion does not require the assumption that all aspects of fashion are designed to transmit messages that enhance social and reproductive success, but only that some features of fashion modify bodily communication in such useful ways. Even this limited claim flies in the face of the common assumption that fashions change purely for the sake of change and have no underlying order apart from the desire of the clothing industry to stimulate sales. Even if some aspects of dress are thus changing quite randomly, it can still be argued that fashions will be more successful if they transmit messages that customers want to send at a particular point in history.

Serious scholars of dress styles have observed regular swings in fashion, including a cycle of dress length going back three centuries, but they have not identified the fundamental drivers of these fashions, probably because they have not looked in the right place. Recent research by the author has found that dress fashions respond to changes in the marriage market and women's economic strategies in much the same way as standards of bodily attractiveness do.

In 1971, Mary Ann Mabry produced her master's dissertation in home economics at the University of Tennessee at Knoxville.[21] She had the rather interesting idea that changes in dress styles, as depicted in women's fashion magazines, might be related to economic growth as measured by changes in stock valuation. Her observation that women's hemlines rise with the Dow Jones Industrial average was the first clear evidence that fashion is linked to economics. This result was dismissed as a statistical fluke because Mabry did not provide any compelling rationale for why the apparently random fluctuations in stock prices should be predicted by the ostensibly random changes in dress fashions. Yet if you take a long view, thinking in terms of years and decades, the stock market has generally been a fairly good predictor of economic growth. One exception was the period following the 1929 crash, in which Americans felt that stocks were an extremely foolish investment, so that stock market averages did not reflect the true economic prospects of the country. Over the past century, rises in stock market valuation have generally been an excellent predictor of gross national product gains, a widely used measure of national wealth.

During economic expansions, more women enter the workforce, and their increased economic power makes them less reliant on marriage as an economic strategy, which weakens the strength of the marriage bond and liberates sexual behavior. By wearing short skirts, women communicate sexual availability.[22] The evidence for this is that the more skin a woman displays in a dating context, the more sexually accessible she is seen to be, according to the reports of American undergraduates.[23] Scantily clad women entering a German nightclub were more likely to be touched by men they encountered in the bar who evidently read their dress style as a romantic invitation.[24]

The fact that skirts become short during years when the stock market is high may reflect the same sort of functional relationship as that between bodily attractiveness and economic growth. Thus, when women are less focused on marriage, they do not have to express feminine modesty in their dress by wearing long skirts to advertise sexual restraint. When economic confidence is high, and the stock market is up, women's occupational prospects are good and this is reflected in their "liberated" dress styles, including short skirts. If long skirts are a symbol of sexual modesty, short skirts are seen to transmit the message that the wearer may be open to the possibility of sexual intimacy outside marriage (which may not be the case).

Why might women ever want to convey such a message? One possibility is that men are attracted to women they see as sexually available and may prefer to date them over women who advertise chastity. Performer Bette Midler captured this dynamic in her quip to the effect that men like women with a history because they hope that history is going to repeat itself with them. Provocative dress may be women's way of competing amongst themselves for desirable men in a difficult marriage market. As such, it may be somewhat deceptive. A miniskirted young women may be far less sexually available than she seems to be. Nevertheless, it is clear that the heyday of the miniskirt, in the 1960s, was also a time when premarital intercourse was both more socially sanctioned and more common than it had been earlier. Today, most women wear jeans and trousers at least some of the time, which may indicate that the double standard of sexual behavior is a thing of the past inasmuch as the dress distinction between men and women has been partially lost.

When economic times are good, women adhere to a slender standard of attractiveness, aspiring to be less stereotypically feminine and sexy, traits considered distinctly negative in most occupations. At the same time, they are communicating sexual availability by means of liberated clothing styles, such as short skirts. These messages seem contradictory, in the sense of simultaneously wanting to turn off and turn on sexual interest, but this complexity on the surface can probably be reduced to economic competition between women both over jobs and over desirable spouses.

Moreover, when femininity is being exaggerated, as in the gigantic bustles worn by Victorian women to amplify their hips and buttocks, sexual restrictedness is simultaneously being communicated by yards of stifling, inconvenient fabric. This fabric forms a symbolic barrier against sexual expression and even has a direct implication of physical restriction in the sense that it impedes movement. These simultaneously transmitted messages of modesty and sexiness only *seem* contradictory. When you consider that Victorian men looked for both sexiness in appearance and sexual modesty in the behavior of wives, the apparent contradiction breaks down into a simple strategy of increasing desirability to potential husbands. Women adopted these styles because they were competing amongst themselves to marry desirable, high-status men.

FASHIONS OF SEXUAL EXPRESSION

Even though free expression of sexual impulses is discouraged and sanctioned in most societies, there is considerable variation in norms of sexual behavior in *different* countries, and in the *same* country at different historical periods and different locations. The most important single factor affecting sexual behavior in women is the extent to which they must rely on husbands for financial support. If women depend principally on economic support from husbands, they express sexual modesty in behavior and in clothing style. They are disinclined to permit sexual intercourse before marriage because this is damaging to their reputation and marriage prospects. In such societies, there is often an extreme and hypocritical emphasis on the importance of virginity for women at the same time that men are allowed to sow their wild oats. This brittle double standard is skillfully depicted in English novels of the eighteenth century, such as *Clarissa*, Samuel Richardson's story of a woman whose reputation is compromised. In these novels, women are depicted as Christmas crackers. Enjoy them once and they lose all value.

In double-standard societies women must be reserved in their sexual expression and reticent in their manner if they are to preserve their good name and reputation. Loss of sexual reputation was a serious matter resulting in complete ostracism, or worse. The social consequences of free sexual expression for women in a double-standard society is skillfully depicted by Tolstoy in his novel *Anna Karenina*, in which the main character has an extramarital affair that becomes common knowledge. The heroine is so distraught by her banishment from polite society that she commits suicide. In some societies of honor, adulterous women are killed, frequently by their own families who feel dishonored by association.

Sexual reserve is typical of societies in which women have good marriage prospects and can make an advantageous marriage only if their sexual reputation is intact. It is thus ironic that they have considerable leverage over the economic resources of men, yet they can best capitalize on their power by seeming obedient and faithful. When women face a weaker marriage market due to scarcity of marriageable men, or when they are less interested in marriage because they can develop independent careers to support themselves and their children, their sexual expression becomes liberated. Women are no longer so concerned about their sexual reputation and are open to the possibility of extramarital sex. Their sexual availability is expressed in dress styles that communicate sexual availability. Thus, the miniskirt appeared at the same time as the sexual revolution in America during the 1960s. Both the fashion of dress and the fashion of sexual behavior in America can be explained as resulting from a rapid drop in the proportion of single young men to single young women.

Emphasizing the important role of economic support by men for sexual restriction of women might seem biased given the unusual level of economic power enjoyed by modern women. It is, nevertheless, quite in keeping with the spirit of the marriage contract through the ages. If we wanted to understand the importance of economic support by husbands for the sexual behavior of women, we could do a thought experiment in which the money provided by husbands for the maintenance of households and support of children was artificially replaced. Since women would no longer need to appeal to prospective husbands by guarding their sexual reputation, their behavior should become more liberated.

This experiment has already been conducted naturally in at least two different ways. One is that, as women's economic clout has increased—with more women entering jobs and professions during the twentieth century—the relative economic importance of husbands has declined. At the same time, women's sexual behavior has become more liberated with a higher percentage of women having sex before marriage, for example.

The Swedish government has inadvertently carried out an even more precise version of this thought experiment. In an effort to encourage more women to seek careers, Sweden developed a welfare state that provided generous subsidies to pregnant women. It paid for child support and all health-care expenses, as well as providing housing expenses, free clothing, and free education. The state thus took on most of the economic roles previously covered by fathers. The result was a steep decline in the institution of marriage and many other social changes, including increased sexual liberation of women and a sharp rise in single-parent families.[25]

Very few Swedish women are virgins when they marry, and virginity is less valued in a marriage partner in Sweden than in any other economically developed country for which this information is available.[26] Divorce rates are also very high, with two-thirds of marriages ending in divorce. These phenomena are at least partly due to the fact that women no longer rely on marriage as an economic strategy for supporting their children and are therefore less concerned about their sexual reputation.

Anthropologists see marriage as a contract in which there is an exchange of economic goods (provided by the husband) for the opportunity to reproduce (provided by the wife). Women who wish to enhance their value as wives must practice sexual restraint because this provides a guarantee of paternity to a potential husband. Where women compete for highly desirable husbands, as is true for upper-caste Indian women, advertisement of good sexual reputation may run to the extreme of locking up virgins in their homes under virtual house arrest by relatives. Such claustration would not work if the young woman felt sexually liberated, of course. In societies of "honor," women learn the social importance of sexual modesty so that they can protect their own sexual reputation.

In summary, where women rely heavily on economic support from husbands to raise their children, premarital chastity is valued. Where male support is less important, women do not need to seem sexually restricted because they are not locked in a competition with other young women to provide a guarantee of paternity to prospective grooms. In Sweden and other sexually liberated countries premarital chastity is not valued. Sexual liberation has broken out.[27]

BEARD FASHIONS AND MARRIAGE

Like women, men are slaves to fashion—but to a lesser degree. One masculine fashion, wearing facial hair, is tied to marriage markets in much the same way as feminine fashion is.

We know far less about male fashions than we do about female ones, for the simple reason that men's appearance has always attracted less attention. Men are evaluated as less physically attractive than women in psychology experiments. Women also take much more care over their physical appearance, possibly because women's desirability as romantic partners is more affected by their appearance than is true for men. That is because many aspects of the romantic desirability of women can be enhanced through careful attention to grooming and dress. Yet it would be a mistake to assume that women are indifferent to the physical appearance of their mates.

The peacock's tail is a costly ornament that evolved to help males of the species attract females. Long considered arbitrary (by analogy with human fashions), the extreme size and elaborate coloration of the peacock's tail plumage are now understood to reflect superior genes. Among humans, both sexes bear ornaments of sexual advertisement, to use Charles Darwin's term. Permanently enlarged breasts in females are unique among primates (most having breast enlargement only while lactating), and recent research has shown that women having larger breasts are healthier, suggesting that this sexually selected trait in humans is also conveying reliable information about health (see chapter 4). Whether the same is true of men's beards and facial hair remains to be seen. The strongest evidence for this view is indirect and concerns the role played by facial hair in male physical attractiveness.

Like the permanent enlargement of women's breasts, men's facial hair has no obvious physiological function and we can assume that it evolved because of its role in communication. Generalizing from the presence of beards and moustaches in other primate species, ornaments that increase the impression of size and ferociousness of these animals, beardedness may serve as a threat display in humans also.[28] Even though many women deny that beards make men attractive (possibly because beards are often worn by unkempt tramps who are dismissed as unsuitable romantic partners), experimental evidence has shown that beardedness generally produces a positive psychological response. Women evaluate bearded men as more physically attractive and as having more of the qualities considered desirable in potential husbands, such as virility and social dominance (see fig. 15).[29] If beards and facial hair in men help to push women's buttons, and there is compelling experimental evidence for this (described above), then it is somewhat of a mystery that so many men would remove this possible advantage by voluntarily shaving. Interestingly, modern women deny—often vehemently—that facial hair is attractive, but as already pointed out, we do not have to accept these responses at face value as the whole story. Neither can we ignore them. Clearly, the modern social environment makes bearded men so socially undesirable for some reason that this outweighs the advantage of its physical attractiveness demonstrated repeatedly in psychology experiments.

This is very odd, but it is not unprecedented. After all, women handicap their own physical attractiveness to men by attempting to be more slender than men want them to be. Similarly, voluptuous women often mute the sexiness of their bodily appearance by wearing loose clothing or carefully tailored "corporate" suits. They have an excellent reason for doing this: consequences for their perceived competence and presumably, their chances of occupational success. Why men handicap themselves by shaving has not been satisfactorily worked out by evolutionary psychologists. Yet it falls into the same kind of pattern. At historical periods when women's standard of attractiveness is curvaceous, as during the Victorian period of the late nineteenth century, men wear beards, sideburns, and moustaches. When women adhere to a slender standard of attractiveness, as at present, men are clean-shaven.

Many of the discoveries about the relationship between the

Fig. 15. This portrait of James A. Garfield, who subsequently became president of the United States, illustrates the contribution of beardedness to impressions of health and manliness. (Reproduced from the Collections of the Library of Congress, LC-B8172-2218)

slender standard and the marriage market also apply to the relationship between shaving and marriage. Study of the undulations of beard fashions in the *Illustrated London News,* between 1842 and 1971 has shown that when women control the marriage market, and when men compete intensely to be married, facial hair fashions are prevalent. These findings were replicated in a study of American beard fashions in the Sears catalog. Men evidently grow beards because this makes it easier for them to attract a wife, just as women exaggerate their bodily curvaceousness at periods when they rely heavily on marriage as an economic strategy. When competition over desirable marriage partners is intense, each sex manipulates the other by exploiting evolved reactions to attractive bodily traits. The problem with this explanation, of course, is that it does not explain why men would ever want to sabotage their own physical attractiveness by shaving.[30]

WHAT WOMEN MAY REALLY DISLIKE
ABOUT MEN'S FACIAL HAIR

Women may accept flamboyant clothing on their favorite musician, but they may dislike it on their boyfriend because they interpret it as a form of flamboyant sexual advertisement; this makes it distinctly unappealing for its implications of sexual promiscuity that most women dislike in a potential husband. Women should be particularly uncomfortable with bearded boyfriends and husbands during historical periods of sexual liberation for women because, at these times, any woman having sex with the boyfriend or husband can be viewed as a serious competitor who may one day marry him.

Men who shave their faces could be muting their sexual signals just as women do, for example when they drape their bodies in the loose gowns favored by nuns, playing down their secondary sexual characteristics. This is a reasonable possibility, but it encounters a snag. Women who dress modestly on one occasion can change their dress or even remove it for another. A man who has shaved has not only reduced his physical attractiveness to a woman's romantic competitors, he has reduced his attractiveness to the woman herself, if the experimental findings on this subject discussed earlier are to be trusted. There must be some other reason that women prefer clean-shaven men when they face a difficult marriage market.

Another possibility is that men may hide their true feelings behind a mask of hair, making it difficult for women to interpret the strength of their romantic interest. Modern women are very concerned about issues of emotional commitment in their boyfriends and spend hours in discussions that revolve around the emotional temperature of their relationships. Men characteristically avoid statements of emotional commitment and may keep their girlfriends in the dark as to the true nature of their romantic feelings and intentions. Some even preserve this noncommital attitude right up to and including the night of their wedding, when many women do not receive the emotional declaration that they had been hoping for.

The deception hypothesis has the advantage of being comparatively easy to test because of the way that emotional expressions are concentrated at the front of the face, around the eyes, the forehead, and the mouth. The point is that facial expressions of emotion are greatly obscured by mustaches that conceal the small movement of muscles at the corners of the mouth which provide the nuances for many emotional expressions such as happiness, sadness, and disgust. This means that when women encounter a difficult marriage market that leaves them open to sexual exploitation, they should prefer men whose emotional transparency is improved by the absence of a mustache. They do not need to be concerned about whether men have goatees or sideburns because these block very little of the emotional expressivity of the face, although they might conceal facial blemishes that are indicative of poor health.

The author tested out the deception hypothesis using sociologist Dwight Robinson's data on the facial hair fashions of men depicted in the *Illustrated London News* between 1841 and 1971. As predicted, mustache fashions declined when women encountered a difficult marriage market (as assessed by the proportion of single men to single women at the peak ages of marriage). Mustaches also disappeared during periods when illegitimacy ratios were high, indicating that women might be particularly vulnerable to (and would thus be concerned about) sexual exploitation. What was most compelling about these results was that mustaches were much more strongly related to the marriage market and to illegitimacy than beards or sideburns were, exactly as predicted by the deception hypothesis because mustaches conceal emotional expressions more than sideburns or beards.[31]

These findings by the author are preliminary and require further verification before being accepted as fact. One of the most enjoyable features of scientific research is being able to make predictions that allow us to make new discoveries. If the deception hypothesis is correct, and if women really avoid men with mustaches because they find them difficult to "read" romantically, then several other phenomena should fall in line. Women who are secure with their current partner should find men with mustaches more attractive than women who feel romantically insecure. Mustache fashions should be more prevalent in societies with stable marriage and low divorce rates than those with unstable marriage. In such societies, women do not have to fear sexual infidelity of husbands because this does not threaten the stability of their marriage, given that the husband's lover is a woman without sexual reputation who cannot be a candidate for marriage. Unmarried women do not have to be unduly concerned about the motives of suitors because during such periods of stable marriage, men typically compete intensely to be married. They use their sexually selected traits, including beards and mustaches, to enhance their desirability as marriage partners.

Why men have beards is no more mysterious than why women have permanently enlarged breasts and narrow waists. All of these traits were produced by sexual selection and are used by prospective mates as indicators of health and biological quality. Far more intriguing is why women should underplay their feminine body shape and why men should handicap their sexual attractiveness by shaving.

The generally positive stereotype produced by bearded men must have some negative connotation that is particularly important when the marriage market for women is weak and they have difficulty marrying a desirable husband. One possibility is that beardedness for men interferes with their professional roles, perhaps by making them seem too unreliable or too individualistic to fit in with job requirements.

Another more interesting speculation is that bearded men are seen as less reliable and less willing to be committed to a single marriage partner. During periods when marriage is unstable, and divorce or desertion are distinct possibilities, women may prefer a man who is less sexually attractive to other women because this increases the likelihood that he will be sexually faithful and that the marriage will last. The interesting wrinkle in this story is that during periods of stable marriage, women do not have to be concerned about their husbands' philandering as a threat to the marriage. During these times, the double standard applies, and men do not marry the women they have sex with since sexually active single women lose their reputation and are considered unsuitable for marriage, as already noted. In more sexually liberated times, premarital intercourse does not have the same negative connotations, and any extramarital affair can be a real threat to the marriage.

A third possibility, examined above, is that where women need to be particularly sensitive to the romantic intentions of men, they prefer partners to be clean-shaven—probably because this allows them to read the emotional signals conveyed by the many muscles around the mouth that change with happiness, sadness, anger, and deception. A man who wears a mustache and beard can be described as wearing a mask, albeit one made of hair. This could make it difficult to assess their level of emotional commitment to a woman. The strongest evidence supporting this hypothesis is that mustache fashions are more strongly connected to the marriage market than facial hair in general. The point is that mustaches conceal the subtle emotional inflections conveyed by scores of tiny muscles around the mouth. Sideburns and beards have similar implications of manliness, and thus allow us to discard any rival explanations having to do with changes in the value attributed by women to stereotypical masculinity.

Whether any of these speculative explanations of shaving turns out to be correct or not, they are clearly the right sort of explanation. Beardedness must have some negative connotation as far as women or potential employers are concerned to promote the current fashion of being clean-shaven. Experiments in social perception have shown that clean-shaven men are seen as more reliable and trustworthy in occupational roles as salesmen and college professors. They may also be seen as more sexually faithful potential husbands because bearded men are seen as more "active," and more "potent," not to mention "dirtier."[32]

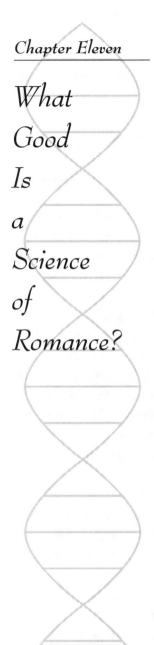

Chapter Eleven

What Good Is a Science of Romance?

Having described many of the fascinating recent findings of evolutionary psychologists and researchers in allied fields, it is time to conclude this book by evaluating the human significance of these findings. The first question to ask is whether we should put the intimate details of our romantic lives under the microscope, either literally, or metaphorically. The second question is related to the first. Even if we accept the premise that romance is fair game for scientific scrutiny, what good are we going to get out of doing this?

IS ROMANCE FAIR GAME FOR SCIENCE?

An evolutionary approach to human sexuality offers a compelling and comprehensive understanding of modern behavior that is based on science and logic. It removes the arbitrariness from modern phenomena, such as varying standards of physical attractiveness. One of the most important contributions of evolutionary psychology is that it helps us to find patterns where none previously existed. Recognizing these patterns is not only important as a way of helping us to understand our own actions and reactions, but it has practical implications.

Thus, the evolutionary approach to sexuality compels us to reinterpret many of the phenomena of our everyday lives. School shootings are no longer just random violence, but a desperate bid for social status on the part of teenagers who have little social status among peers and who have little appeal to the opposite sex. To reinterpret high-school shooting as mating aggression may seem unhelpful to many people, particularly if they happen to know the victims, but it is helpful to know that the shootings were not entirely random. In a number of cases, they occurred in an identifiable social context and the perpetrators had often identified themselves as at risk for engaging in such behavior. Their self-identified motivation was to raise their social status and escape ridicule and bullying. This falls into an evolutionary pattern of young men engaging in risky behavior to raise their social status, a pattern that also contributes to "road rage," another bane of society.

Why is it that when a person is cut off on the highway there is such a tendency to react with rage? The encounter ought not to matter any more than hitting a nasty pothole, but we react as though our whole future was on the line, as though we are going to lose status and respect. Because our ancestors lived in small groups whose members were well known to each other, natural selection has designed our brains to deal with people we interact with repeatedly; it has not equipped us to cope with anonymous strangers in single encounters. In the distant past, losing face in any one incident would have been to suffer a lasting loss of respect and social power. Our emotional reactions do not allow us to discriminate between the anonymous face-off on the highway—in which an unknown roadhog briefly crosses our path and may never be seen again—and a serious loss of reputation among people who know us well and are arbiters of our social status.

We know that men, and particularly young ones, are most hot-headed about their reputation because this would have had serious reproductive consequences in the evolutionary past, but there is no reason that women cannot have a similar reaction to the frustration of being bullied on the highway. Women are increasingly being implicated in road-rage episodes. This could be because modern women have higher economic power and occupational status than in the past, so that they are more accustomed to responding assertively in situations where prestige is on the line.

The modern environment makes people's romantic behavior very different from what it would have been in the evolutionary past, and evolutionary psychologists have begun to unravel what happens when human Darwinian adaptations collide with the modern environment. Modern evolutionary psychology shows that some aspects of human nature are close to being universal, such as interpretation of physical beauty; emotional responses to the loss of a child, friend, or lover; or the expression of facial emotion. It is important to understand these aspects of our evolutionary heritage. It is more interesting to know why people in different societies behave so differently than it is to understand their universal similarities. What is really fascinating about human romantic behavior is to see how our hereditary adaptations are affected by the modern environment.

What does this mean? Take the example of beards. Darwin recognized that the human beard is a sexual ornament that helps men to attract women. This universalist explanation runs into some intriguing problems when we study beards around the world. The first is that men in several populations have almost no facial hair, perhaps because skin quality of the face was useful to women in selecting a mate. This would be true, for example, in tropical climates where vulnerability to diseases such as smallpox shows up as facial pockmarks. The second problem is that where modern men are biologically endowed with luxuriant facial hair, they may go to great lengths to remove it.

Superficially, the habit of shaving makes no sense to an evolutionist. Yet what looks like an absurdity turns out to have an unexpected and finely wrought order. Shaving fashion corresponds with social environments in which women have good reason to be concerned about the level of romantic commitment of their partners.

The kind of people we become is also very much shaped by our social environment, although this seemingly unexceptional view has been attacked by Judith Harris, who persists in seeing nature and nurture as irreconcilable opposites. Thus, the development of a person's romantic interests are very much a feature of their childhood. Children who are raised in stressful social environments are much more likely to experience high levels of conflict and subsequent instability in their romantic lives.

We are sometimes surprised to discover that successful actors and actresses have had difficult childhoods in which parents fought a great deal or were physically separated. This is surprising because such a background does not predict success in other fields. Material success can be accompanied by a painful personal life of emotional insecurity and self-doubt, as exemplified by movie stars such as Frances Farmer and Marilyn Monroe. A rocky history of successive marriages, divorces, and remarriages is such a norm of Hollywood that you would think stars organized their whole lives for the convenience of tabloid newspapers. They are addicted to being in love, as well as to available chemical substitutes, but the chemistry of their addiction inevitably lets them down.

All of these features of the stereotypical Hollywood star are consistent with a difficult early environment, setting them up for troubled romances and unstable marriages. The only real mystery is why they should turn out to be such talented performers. Even this puzzle makes some kind of sense because children who suffer from feelings of emotional neglect have an unusual thirst for the emotional gratification and sense of importance that comes from impressing an audience.

It is interesting that the romantic lives of individuals are very much influenced by changes in societies over time. Thus, the relations between husbands and wives are influenced by changes in the perceived value of marriage for men and women. Throughout most of recorded history, women have had a strong desire to marry because they relied heavily on the economic contribution of husbands to enable them to raise their children. When the economic motive is taken away or muted, women are less likely to marry and their marriages are more likely to be unstable.

These points are nicely illustrated by the case of Sweden, where the welfare state promoted employment among women by increasing supports for children, allowing single mothers to raise children in relative comfort without the help of husbands. The result is that marriage rates are among the lowest in the world. For those who do marry, two-thirds of their unions end in divorce, which is one of the highest dissolution rates in the developed world.

When women have trouble marrying—as happened in the 1960s in America, but was also characteristic of Chaucer's England, and of Sparta in the age of Aristotle—they become sexually liberated. "Sexual liberation" is probably an unfortunate term because such periods actually favor the romantic agendas of men and can be hostile to the evolved emotional needs of women. Thus, in the current social environment, many women complain of a lack of emotional commitment from their lovers and settle for temporary cohabitation arrangements when they would prefer permanent marriages. In more sexually prudish times, women had to be on guard to protect their sexual reputations because a history of premarital intercourse could severely damage their marriage prospects. Ironically, given their sexual repression, they had real power in the marriage market and could negotiate a lifelong contract of economic support for themselves and their children.

Study of existing hunter-gatherer communities has suggested to anthropologists that sexual liberation, combined with greater economic power for women, may have been more typical of our ancestors and that the agricultural societies that make up much of the record of history were the real anomaly. In agricultural societies, men control most of the property and the means of production, and they use their economic power to control women's fertility by stipulating that a bride must be a virgin, for example. Women who make this commitment to exclusive paternity of the husband are entitled to a huge economic commitment from the husband in return.

Evolutionary thinking helps us to understand historical and social changes as a matter of science rather than as a function of the moralistic formulae that were previously employed by scholars. This is a major step in our enlightenment, analogous to the discovery that mental illnesses were produced by brain disorders rather than being the moral defects (or even demon possessions) that scholars had previously assumed them to be. Such enlightenment is clearly important in helping us to accept the social diversity we encounter in the different habits of people in societies around the world, not to mention within our own society. It should also help us to solve practical problems.

Some of the important applications of the evolutionary science of romance have to do with the differing perceptions of men and women of a given situation. Thus, if you want to understand why a woman might perceive the persistent requests of one of her coworkers for a date as sweet and romantic, while exactly the same actions by another coworker can precipitate a lawsuit against the company, you cannot make any progress unless you begin with evolved mate selection criteria. Women are less likely to mention harassment when the man is a desirable candidate for romance than when he is not. These criteria are largely a product of the evolutionary history of our species.

Another very practical application of the science of romance has to do with the emergence of eating disorders as a serious health problem during the 1920s and again from the 1960s on. The novel behavior of young women starving themselves to dangerous extremes makes sense only in terms of their desire to play down evolved sexual signals that interfere with their ability to appear professionally competent. To an evolutionary biologist, this is an almost incomprehensible phenomenon, given the strength of selection pressures favoring food consumption and survival. Yet this recent phenomenon can be understood in terms of the collision of evolved signaling mechanisms with the unique modern environment. The same general point can be made about such diverse—but distinctively modern phenomena—as high school shootings, the causes of divorce, and the troubled romantic lives of Hollywood stars. None of these explanations needs to be repeated.

As an evolved species on this planet, we are on the brink of a new level of scientific self-understanding as far as the mysteries of our sexual relationships are concerned. Thus, evolutionary insights can help us to solve conflict in romantic relationships by recognizing that they often arise from differing reproductive strategies between men and women. The more scientific effort that is applied to relations between the sexes, the better our prospects of resolving conflicts between men and women. Our romantic lives are illuminated by scientific scrutiny, and this understanding not only helps us to appreciate our unique place in nature but also solves practical problems.

Notes

CHAPTER ONE: THE SEXUAL BRAIN

1. J. Kalat, *Biological Psychology* (Pacific Grove, Calif.: Brooks/Cole, 1995).

2. D. Kimura, *Sex and Cognition* (Cambridge: MIT Press, 1999).

3. D. C. Geary, *Male, Female: The Evolution of Human Sex Differences* (Washington, D.C.: American Psychological Association, 1998). Also, Kimura, *Sex and Cognition*; D. Kimura and E. Hampson, "Neural and Hormonal Mechanisms Mediating Sex Differences in Cognition," in *Biological Approaches to the Study of Human Intelligence*, ed. P. A. Vernon (Norwood, N.J.: Ablex, 1992); L. Mealey, *Sex Difference: Developmental and Evolutionary Strategies* (San Diego: Academic Press, 2000).

4. I. Silverman and M. Eals, "Sex Differences in Mental Abilities," in *The Adapted Mind*, ed. J. Barkow, L. Cosmides, and J. Tooby (Oxford: Oxford University Press, 1992), pp. 533–49.

5. Kimura, *Sex and Cognition*.

6. T. Ebert et al., "Increased Cortical Representation of the Fingers of the Left Hand in String Players," *Science* 270 (1995): 305–307.

7. D. Kimura, "Sex Differences in the Brain," *Scientific American* (1992): 119–25.

8. G. P. Murdock, *Culture and Society* (Pittsburgh, Pa.: University of Pittsburgh Press, 1965).

9. Kimura, *Sex and Cognition*.

10. J. Money and A. A. Ehrhard, *Man and Woman, Boy and Girl* (Baltimore, Md.: Johns Hopkins University, 1972); J. Harris, *The Nurture Assumption* (New York: Free Press, 1998), p. 222.

11. Ibid.

12. D. Freeman, *The Fateful Hoaxing of Margaret Mead: A Historical Analysis of Her Samoan Research* (Boulder, Colo.: Westview, 1999).

13. Ibid.

14. J. Kalat, *Biological Psychology*, 5th ed. (Pacific Grove, Calif.: Brooks/Cole, 1995).

15. Geary, *Male, Female*.

16. Kalat, *Biological Psychology*.

17. Geary, *Male, Female*.

18. A. A. Ehrhardt, "Sexual Orientation after Prenatal Exposure to Exogenous Estrogen," *Archives of Sexual Behavior* 14 (1985): 57–77.

19. Kimura, *Sex and Cognition*.

20. S. LeVay, "A Difference in Hypothalamic Structure between Heterosexual and Homosexual Men," *Science* 253 (1991): 1034–37.

21. Kalat, *Biological Psychology*.

22. Kimura, *Sex and Cognition*.

23. K. J. Pucker et al., "Physical Attractiveness of Boys with Gender Identity Disorder," *Archives of Sexual Behavior* 22 (1993): 23–36.

24. R. E. Franken, *Human Motivation* (Belmont, Calif.: Wadsworth, 1994).

25. H. Fisher, *Anatomy of Love* (New York: Norton, 1992).

26. Kalat, *Biological Psychology*.

27. A. C. DeVries et al., "The Effects of Stress on Social Preferences Are Sexually Dimorphic in Prairie Voles," *Proceedings of the National Academy of Sciences of the United States* 93 (1996): 11980–84.

28. S. S. Brehm, and S. M. Kassin, *Social Psychology* (Boston: Houghton Mifflin, 1990).

29. E. T. Ben-Ari, "Pheromones: What's in a Name?" *BioScience* 48, no. 7 (1998): 505–11.

30. B. W. Cutler, "Pheromonal Influences on Sociosexual Behavior in Men," *Archives of Sexual Behavior* 27 (1998): 1–13.

CHAPTER TWO: PHYSICAL ATTRACTIVENESS AND SEX SIGNALS

1. R. Thornhill and S. W. Gangestad, "Human Facial Beauty: Averageness, Symmetry, and Parasite Resistance," *Human Nature* 4 (1993): 237–69.

2. N. Barber, "The Evolutionary Psychology of Physical Attractiveness: Sexual Selection and Human Morphology," *Ethology and Sociobiology* 16 (1995): 395–424; L. Jackson, *Physical Appearance and Gender* (Albany: State University of New York Press, 1992).

3. Barber, "The Evolutionary Psychology of Physical Attractiveness."

4. J. A. Alcock, *Animal Behavior* (Sunderland, Mass.: Sinauer, 1989).

5. Barber, "The Evolutionary Psychology of Physical Attractiveness."

6. W. D. Hamilton and M. Zuk, "Heritable True Fitness and Bright Birds: A Role for Parasites," *Science* 218 (1982): 384–87.

7. M. Petrie, "Increased Growth and Survival of Offspring of Peacocks with More Elaborate Trains," *Nature* 371 (1994): 598–99.

8. C. Darwin, *The Descent of Man and Selection in Relation to Sex* (London: Murray, 1871).

9. Jackson, *Physical Appearance and Gender.*

10. D. Symons, *The Evolution of Human Sexuality* (New York: Oxford University Press, 1979).

11. Ibid.

12. Ibid.

13. D. C. Geary, *Male, Female: The Evolution of Human Sex Differences* (Washington, D.C.: American Psychological Association, 1998).

14. J. R. Udry and B. K. Eckland, "Benefits of Being Attractive: Differential Payoffs for Men and Women," *Psychological Reports* 54 (1984): 47–56.

15. Jackson, *Physical Appearance and Gender.*

16. E. V. Walster et al., "The Importance of Physical Attractiveness in Dating Behavior," *Journal of Personality and Social Psychology* 4 (1966): 508–16.

17. J. Townsend, *What Women Want—What Men Want* (New York: Oxford University Press, 1998).

18. Barber, "The Evolutionary Psychology of Physical Attractiveness."

19. S. Feinman and G. W. Gill, "Females' Response to Male's Beardedness," *Perceptual and Motor Skills* 58 (1977): 533–34.

20. R. J. Pellegrini, "Impressions of the Male Personality as a Functioned of Beardedness," *Psychology* 10 (1973).

21. E. Hatfield and S. Sprecher, *Mirror, Mirror: The Importance of Looks in Everyday Life* (Albany: State University of New York Press, 1986).

22. W. E. Addison, "Beardedness as a Factor in Perceived Masculinity," *Perceptual and Motor Skills* 68 (1989): 921–22.

23. J. A. Reed and E. M. Blunk, "The Influence of Facial Hair on Impression Formation," *Social Behavior and Personality* 18 (1990): 169–76.

24. A. Hellstroem and J. Tekle, "Person Perception through Facial Photographs: Effects of Glasses, Hair, and Beard, on Judgments of Occupation and Personal Qualities," *European Journal of Social Psychology* 24 (1994): 693–705.

25. N. Barber, "Mustache Fashion Covaries with a Good Marriage Marker for Women," *Journal of Nonverbal Behavior* 25 (2001): 261–72.

26. Jackson, *Physical Appearance and Gender.* Also, M. Argyle, *The Psychology of Social Class* (London: Routledge, 1994).

27. J. A. Sheppard and J. Strathman, "Attractiveness and Height: The Role of Stature in Dating Preference, Frequency of Dating, and Perceptions of Physical Attractiveness," *Personality and Social Psychology Bulletin* 8 (1989): 53–58. Also, Jackson, *Physical Appearance and Gender.*

28. Hatfield and Sprecher, *Mirror, Mirror.* Also, Jackson, *Physical Appearance and Gender;* Barber, "The Evolutionary Psychology of Physical Attractiveness."

29. T. Horvath, "Correlates of Physical Beauty in Men and Women," *Social Behavior and Personality* 7 (1974): 145–51.

30. Thornhill and Gangestad, "Human Facial Beauty."

31. A. P. Moller, "Female Swallow Preference for Symmetrical Male Sexual Ornaments," *Nature* 357 (1992): 238–40.

32. R. Thornhill and S. W. Gangestad, "Human Fluctuating Asymmetry and Sexual Behavior," *Psychological Science* 5 (1994): 297–302. Also, R. Thornhill and S. W. Gangestad, "Human Female Orgasm and Mate Fluctuating Asymmetry," *Animal Behavior* 50 (1995): 1601–15.

33. Barber, "The Evolutionary Psychology of Physical Attractiveness." Also, Jackson, *Physical Appearance and Gender*.

34. Barber, "The Evolutionary Psychology of Physical Attractiveness."

35. Jackson, *Physical Appearance and Gender*.

36. D. Singh, "Adaptive Significance of Female Physical Attractiveness: Role of Waist-to-Hip Ratio," *Journal of Personality and Social Psychology* 65 (1993): 293–307.

37. Barber, "The Evolutionary Psychology of Physical Attractiveness."

38. Ibid.

39. J. T. Manning et al., "Breast Asymmetry and Phenotypic Quality in Women," *Ethology and Sociobiology* 18 (1997): 223–36. Also Barber, "The Evolutionary Psychology of Physical Attractiveness."

40. Jackson, *Physical Appearance and Gender*.

41. D. I. Perett, K. A. May, and S. Yoshikawa, "Facial Shape and Judgments of Female Attractiveness," *Nature* 368 (1994): 239–42.

42. Barber, "The Evolutionary Psychology of Physical Attractiveness."

43. Ibid.

44. V. S. Johnson and M. Franklin, "Is Beauty in the Eye of the Beholder?" *Ethology and Sociobiology* 14 (1993): 183–99.

Chapter Three: Love's Labors: Dating Competition and Aggression

1. J. Blank and C. Warren, "Prayer Circle Murders," *U.S. News & World Report,* 15 December 1997, p. 276.

2. P. D. Maclean, "The Triune Brain in Conflict," *Psychotherapy and Psychosomatics* 28 (1977): 207–20.

3. M. Daly and M. Wilson *Sex, Evolution, and Behavior*, 2d ed. (Belmont, Calif.: Wadsworth, 1983).

4. Ibid.

5. T. T. Jackson and G. McCauley, "Field Study of Risk-Taking

Behavior of Automobile Drivers," *Perceptual and Motor Skills* 43 (1976): 471–74.

6. Daly and Wilson, *Sex, Evolution, and Behavior*.

7. M. Wilson and M. Daly, "Competitiveness, Risk-Taking, and Violence: The Young Male Syndrome," *Ethology and Sociobiology* 6 (1985): 59–73.

8. A. Booth and J. M. Dabbs, "Testosterone and Men's Marriages," *Social Forces* 72 (1993): 463–77.

9. H. Nyborg, *Hormones, Sex, and Society* (Westport, Conn.: Praeger, 1994). Also, J. M. Dabbs Jr., R. Strong, and R. Milun, "Explaining the Mind of Testosterone: A Beeper Study," *Journal of Research in Personality* 31 (1997): 577–87.

10. G. F. Miller, in *The Evolution of Culture*, ed. R. Dunbar, C. Knight, and C. Power (Edinburgh: University of Edinburgh Press, in press).

11. M. Virkunen et al., "Relationship of Psychobiological Variables to Recidivism in Violent Offenders and Impulsive Fire-Setters," *Archives of General Psychiatry* 46 (1989): 600–603. Also, J. Kalat, *Biological Psychology*, 5th ed. (Pacific Grove, Calif.: Brooks/Cole, 1995).

12. D. Lykken, *The Antisocial Personalities* (Hillsdale, N.J.: Lawrence Erlbaum, 1995).

13. D. C. Geary, *Male, Female: The Evolution of Human Sex Differences* (Washington, D.C.: American Psychological Association, 1998).

14. D. Kimura, *Sex and Cognition* (Cambridge: MIT Press, 1999).

15. B. Russell, *A History of Western Philosophy* (New York: Simon and Schuster, 1972).

16. K. Lorenz, *On Aggression* (New York: Bantam Books, 1980).

17. E. Ebbeson, B. Duncan, and V. Konecni, "Effects of Content of Verbal Aggression on Future Verbal Aggression: A Field Experiment," *Journal of Experimental Psychology* 11 (1975): 192–204.

18. A. Bandura, *Aggression: A Social Learning Analysis* (Englewood Cliffs, N.J.: Prentice Hall, 1973).

19. F. B. Steuer, J. M. Applefield, and R. Smith, "Televised Violence and the Interpersonal Aggression of Preschool Children," *Journal of Experimental Child Psychology* 11 (1971): 422–47.

20. L. P. Eron et al., "Does Television Violence Cause Aggression?" *American Psychologist* 27 (1972): 253–63. Also, L. R. Huesman, "Psychological Processes Promoting the Relation Between Exposure to Media

Violence and Aggressive Behavior by the Viewer," *Journal of Social Issues* 42 (1986): 125–39.

21. J. Leo, "When Life Imitates Video," *U.S. News & World Report,* 3 May 1999, p. 14.

22. G. M. B. Hanson, "The Violent World of Video Games," *Insight on the News,* 28 June 1999, p. 44.

23. Ibid.

24. A. Cannon et al., "Why?" *U.S. News & World Report,* 3 May 1999, p. 16.

25. Ibid.

26. Ibid.

27. J. Blank, "The Kid No One Noticed: Guns He Concluded Could Get His Classmates' Attention," *U.S. News & World Report,* 12 October 1998, p. 27.

28. Ibid.

29. Ibid.

30. Ibid.

CHAPTER FOUR: THE DANCE OF THE SEXES

1. K. Grammar, "Human Courtship Behavior: Biological Basis and Cognitive Processing," in *The Sociobiology of Sexual and Reproductive Strategies,* ed. A. E. Rasa, C. Vogel, and E. Voland (London: Chapman and Hall, 1989), pp. 147–69.

2. T. Perper, *Sex Signals: The Biology of Love* (Philadelphia: ISI Press, 1985).

3. D. C. Geary, *Male, Female: The Evolution of Human Sex Differences* (Washington, D.C.: American Psychological Association, 1998).

4. R. D. Alexander and K. M. Noonan, "Concealment of Ovulation, Parental Care, and Social Evolution," in *Evolutionary Biology and Human Social Behavior,* ed. R. D. Alexander and K. M. Noonan (North Scituate, Mass.: Duxbury Press, 1979), pp. 402–35.

5. M. Daly and M. Wilson, *Sex, Evolution, and Behavior,* 2d ed. (Belmont, Calif.: Wadsworth, 1983).

6. Alexander and Noonan, "Concealment of Ovulation, Parental Care, and Social Evolution."

7. S. S. Brehm and S. M. Kassin, *Social Psychology* (Boston: Houghton Mifflin, 1990).

8. D. Kimura, *Sex and Cognition* (Cambridge: MIT Press, 1999).

9. Geary, *Male, Female.*

10. J. Townsend, *What Women Want—What Men Want* (New York: Oxford University Press, 1998).

11. Ibid.

12. Ibid.

13. D. M. Buss, *The Evolution of Desire: Strategies of Human Mating* (New York: Basic Books, 1994).

14. Townsend, *What Women Want—What Men Want.*

15. Ibid.

16. Ibid.

17. D. Symons, *The Evolution of Human Sexuality* (New York: Oxford University Press, 1979).

18. Buss, *The Evolution of Desire.*

19. A. C. Kinsey, W. B. Pomeroy, and C. E. Martin, *Sexual Behavior in the Human Male* (Philadelphia: W. B. Saunders, 1948). Also, A. C. Kinsey et al., *Sexual Behavior in the Human Female* (Philadelphia: W. B. Saunders, 1953); J. H. Jones, *Alfred C. Kinsey: A Public/Private Life* (New York: Norton, 1997).

20. W. H. Masters and V. E. Johnson, *Human Sexual Response* (Boston: Little Brown, 1966).

21. C. F. Westoff, R. G. Potter, and P. C. Sagi, *The Third Child* (Princeton: Princeton University Press, 1963).

22. J. F. Saiucier, "Correlates of the Long Postpartum Taboo: A Cross-Cultural Study," *Current Anthropology* 13 (1972): 238–49.

23. R. S. Baker and M. A. Bellis, *Human Sperm Competition* (London: Chapman and Hall, 1995).

CHAPTER FIVE: THE CHEATING HEARTS OF BIRDS AND HUMANS

1. N. B. Davies, "Polyandry, Cloaca-Pecking, and Sperm Competition in Dunnocks," *Nature* 302 (1983): 334–36.

2. A. Jolly, *The Evolution of Primate Behavior* (New York: MacMillan, 1985).

3. D. M. Buss, *The Evolution of Desire: Strategies of Human Mating* (New York: Basic Books, 1994).

4. D. Symons, *The Evolution of Human Sexuality* (New York: Oxford University Press, 1979).

5. J. Alcock, *Animal Behavior: An Evolutionary Approach* (Sunderland, Mass.: Sinauer, 1988). Also, T. Birkhead, *Promiscuity: An Evolutionary History of Sperm Competition* (Cambridge, Mass.: Harvard University Press, 2000); R. S. Baker and M. A. Bellis, *Human Sperm Competition* (London: Chapman and Hall, 1995).

6. Birkhead, *Promiscuity.* Also, D. C. Geary, *Male, Female: The Evolution of Human Sex Differences* (Washington, D.C.: American Psychological Association, 1998).

7. Baker and Bellis, *Human Sperm Competition.*

8. M. Daly and M. Wilson, *Homicide* (Hawthorne, N.Y.: Aldine de Gruyter, 1995).

9. Buss, *The Evolution of Desire.*

10. P. G. Schnitzer and C. W. Runyan, "Injuries to Women in the United States: An Overview," *Women's Health* 23 (1995): 9–27. Also, P. W. Cook, *Abused Men* (Westport, Conn.: Greenwood, 1997).

11. Buss, *The Evolution of Desire.*

12. A. D. La Violetta and O. W. Barnett, *It Could Happen to Anyone: Why Battered Women Stay* (Beverly Hills, Calif.: Sage, 2000).

13. M. Daly and M. Wilson, *Sex, Evolution, and Behavior*, 2d ed. (Belmont, Calif.: Wadsworth, 1983).

14. Buss, *The Evolution of Desire.*

15. D. M. Buss et al., "Sex Differences in Jealousy: Evolution, Physiology, and Psychology," *Psychological Science* 3 (1992): 251–55.

16. A. J. Bateman, "Intrasexual Selection in Drosophila," *Heredity* 2 (1948): 349–68.

17. R. Thornhill, "Panorpa (*Mecoptera Panorpidae*) Scorpionflies: Systems for Understanding Resource-Defense Polygyny and Alternative Male Reproductive Efforts," *Annual Review of Ecology and Systematics* 12 (1981): 355–86.

18. Symons, *The Evolution of Human Sexuality.*

19. M. V. Studd, "Sex, Power, and Conflict in Sexual Harassment

Cases: Evolutionary and Feminist Perspectives" (paper presented to the Evolution and Human Behavior Society, Ann Arbor, Mich., June 1994).

20. Ibid.

21. Thornhill, "Panorpa Scorpionflies."

22. Jolly, *The Evolution of Primate Behavior.*

23. R. Thornhill and C. T. Palmer, *A Natural History of Rape: Biological Bases of Sexual Coercion* (Cambridge: MIT Press, 2000).

24. S. Brownmiller, *Against Our Will: Men, Women, and Rape* (New York: Fawcett, 1993).

25. N. W. Thornhill and R. Thornhill, "Human Rape: An Evolutionary Analysis," *Ethology and Sociobiology* 4 (1983): 137–73. Also, Symons, *The Evolution of Human Sexuality*; M. J. Hindelag and B. J. Davis, "Forcible Rape in the United States: A Statistical Profile," in *Forcible Rape: The Crime, the Victim, and the Offender*, ed. M. J. Hindelag and B. J. Davis (New York: Columbia University Press, 1977), pp. 87–114.

26. Ibid. Also, M. Amir, *Patterns in Forcible Rape* (Chicago: Chicago University Press, 1971); J. M. MacDonald, *Rape Offenders and Their Victims* (Springfield, Ill.: C. C. Thomas, 1971).

27. Symons, *The Evolution of Human Sexuality.*

28. Ibid.

29. M. E. P. Seligman, *What You Can Change and What You Can't* (New York: Fawcett Columbine, 1993).

30. Baker and Bellis, *Human Sperm Competition.*

31. J. Townsend, *What Women Want—What Men Want* (New York: Oxford University Press, 1998).

32. Symons, *The Evolution of Human Sexuality.*

33. Seligman, *What You Can Change.*

34. S. Rachman and J. Hodgson, "Experimentally Induced 'Sexual Fetishism': Replication and Development," *Psychological Record* 18 (1968): 25–27.

35. Ibid.

CHAPTER SIX: WHY MARRIAGES FAIL

1. D. M. Buss, *The Evolution of Desire: Strategies of Human Mating* (New York: Basic Books, 1994).

2. Department of Health and Human Services, *Vital Statistics of the United States,* vol. 3, *Marriage and Divorce* (Washington, D.C.: GPO, 1995).

3. B. R. Mitchell, *British Historical Statistics* (New York: Cambridge University Press, 1988).

4. M. Guttentag and P. F. Secord, *Too Many Women: The Sex Ratio Question* (Beverly Hills, Calif.: Sage, 1983).

5. D. Symons, *The Evolution of Human Sexuality* (New York: Oxford University Press, 1979).

6. M. Daly and M. Wilson, *Sex, Evolution, and Behavior,* 2d ed. (Belmont, Calif.: Wadsworth, 1983). Also, C. Chelala, "An Alternative Way to Stop Female Genital Mutilation," *Lancet* 352 (1998): 126.

7. Buss, *The Evolution of Desire.*

8. Department of Health and Human Services, *Vital Statistics of the United States.*

9. Guttentag and Secord, *Too Many Women.*

10. Symons, *The Evolution of Human Sexuality.* Also, Guttentag and Secord, *Too Many Women.*

11. M. Ember, "Alternative Predictors of Polygyny," *Behavioral Science Research* 19 (1985): 1–23. Also, B. Strassman, "Polygyny as a Risk Factor for Child Mortality Among the Dogon," *Current Anthropology* 38 (1997): 688–95; Symons, *The Evolution of Human Sexuality.*

12. J. Townsend, *What Women Want—What Men Want* (New York: Oxford University Press, 1998).

13. Guttentag and Secord, *Too Many Women.*

14. N. Barber, *Why Parents Matter: Parental Investment and Child Outcomes* (Westport, Conn.: Bergin and Garvey, 2000).

15. Townsend, *What Women Want—What Men Want.*

16. Guttentag and Secord, *Too Many Women.*

17. D. Hamer et al., "A Linkage Between DNA Markers on the X Chromosome and Male Sexual Orientation," *Science* 261 (1993): 321–27.

18. "Stanford White and Evelyn Nesbit," *People,* 12 February 1996, pp. 79–80.

19. Buss, *The Evolution of Desire.*

20. A. O'Neill, "Webstruck: A Lonely Wife Searches for Love on the Internet Leading to Her Murder at Home," *People,* 2 June 1997, pp. 85–87. Also, E. Rivera, "Blood Roses: A Correspondence on the Internet Leads to Carnage," *Time,* 10 February 1997, p. 40.

21. Buss, *The Evolution of Desire.*

22. R. D. Alexander, *Darwinism and Human Affairs* (Seattle: University of Washington Press, 1979).

Chapter Seven: Learning to Love

1. F. L. Guiles, *Legend: The Life and Death of Marilyn Monroe* (Lanham, Md.: Scarborough House, 1991). Also, F. J. Griffin, *If You Can't Be Free, Be a Mystery: In Search of Billie Holiday* (New York: Free Press, 2001).

2. J. R. Harris, *The Nurture Assumption: Why Children Turn Out the Way They Do* (New York: Free Press, 1998).

3. N. Barber, *Why Parents Matter: Parental Investment and Child Outcomes* (Westport, Conn.: Bergin and Garvey, 2000).

4. R. L. Trivers, "Parental Investment and Sexual Selection," in *Sexual Selection and the Descent of Man*, ed. B. Campbell (Chicago, Ill.: Aldine-Atherton, 1972), pp. 136–79.

5. Barber, *Why Parents Matter.*

6. S. Provence and R. C. Lipton, *Infants in Institutions* (New York: International Universities Press, 1962).

7. L. Malsen, *Wolf Children and the Problem of Human Nature* (New York: Monthly Review Press, 1972).

8. C. Blakemore, "Developmental Factors in the Formation of Feature-Extracting Neurons," in *The Neurosciences: Third Study Program*, ed. F. G. Worden and F. O. Schmitt (Cambridge: MIT Press, 1974). Also, Barber, *Why Parents Matter.*

9. L. Yarrow, "Maternal Deprivation: Toward an Empirical and Conceptual Reevaluation," *Psychological Bulletin* 58 (1961): 459–90.

10. M. R. Rosenzweig, E. L. Bennett, and M. C. Diamond, "Brain Changes in Response to Experience," *Scientific American* 226, no. 2 (1972): 22–29.

11. C. T. Ramey, "High Risk Children and IQ: Altering Intergenerational Patterns," *Intelligence* 16 (1992): 239–56.

12. P. Vorria et al., "A Comparative Study of Greek Children in Long-Term Residential Group Care and in Two-Parent Families," *Journal of Child Psychology and Psychiatry and Allied Disciplines* 39 (1998): 225–36.

13. Barber, *Why Parents Matter*.

14. M. D. S. Ainsworth et al., *Patterns of Attachment: A Psychological Study of the Strange Situation* (Hillsdale, N.J.: Erlbaum, 1978).

15. Ibid.

16. J. Belsky, L. Steinberg, and P. Draper, "Childhood Experience, Interpersonal Development, and Reproductive Strategy: An Evolutionary Theory of Socialization," *Child Development* 62 (1991): 647–70.

17. M. D. S. Ainsworth, "Attachment as Related to Mother-Infant Interaction," in *Advances in the Study of Behavior,* vol. 9, ed. J. S. Rosenblatt et al. (Orlando, Fla.: Academic Press, 1979).

18. D. C. van den Boom, "Preventive Intervention and the Quality of Mother-Infant Interaction and Infant Exploration in Irritable Infants," in *Psychology Behind the Dikes*, ed. W. Koops et al. (Amsterdam: Eburon, 1990).

19. L. Youngblade and J. Belsky, "Child Maltreatment, Infant-Parent Attachment Security, and Dysfunctional Peer Relations in Toddlerhood," *Topics in Early Child Special Education* 9 (1989): 1–15. Also, H. R. Starr, "The Lasting Effects of Child Maltreatment," in *Human Development*, 22d ed., ed. L. Fenson and J. Fenson (Guildford, Conn.: Dushkin, 1994), pp. 142–46.

20. P. Hazan and P. R. Shaver, "Romantic Love Conceptualized as an Attachment Process," *Journal of Personality and Social Psychology* 52 (1987): 511–24. Also, N. Barber, "Sex Differences in Disposition Towards Kin, Security of Adult Attachment, and Sociosexuality as a Function of Parental Divorce," *Evolution and Human Behavior* 19 (1998): 1–8; E. M. Hill, J. P. Young, and J. L. Nord, "Childhood Adversity, Attachment Security, and Adult Relationships: A Preliminary Study," *Ethology and Sociobiology* 15 (1994): 323–38.

21. G. R. Patterson, B. D. DeBaryshe, and E. Ramey, "A Developmental Perspective on Antisocial Behavior," *American Psychologist* 44 (1989): 329–35.

22. Ibid.

23. P. Long et al., "Does Parent Training with Young Noncompliant Children Have Long-Term Effects?" *Behavioral Research and Therapy* 32 (1994): 101–107. Also, Barber, *Why Parents Matter*.

24. N. Barber, *Parenting: Roles, Styles, and Outcomes* (Commack, N.Y.: Nova Science, 1998).

25. B. Hart and T. Risley, *Meaningful Differences in the Everyday Experience of Young American Children* (Baltimore, Md.: Paul H. Brookes, 1995).

26. J. S. Musick, *Young, Poor, and Pregnant: The Psychology of Teenage Motherhood* (New Haven, Conn.: Yale University Press, 1993).

27. M. Guttentag and P. F. Secord, *Too Many Women: The Sex Ratio Question* (Beverly Hills, Calif.: Sage, 1983).

28. E. Anderson, *Streetwise* (Chicago: University of Chicago Press, 1990).

CHAPTER EIGHT: SEX AND WORK

1. G. P. Murdock, *Culture and Society* (Pittsburgh, Pa.: University of Pittsburgh Press, 1965).

2. "Home Front Heroine," *People* 47, 16 June 1997, p. 118.

3. R. Lynn, "Sex Differences in Competitiveness and the Valuation of Money in Twenty Countries," *Journal of Social Psychology* 133 (1993): 507–11.

4. A. Konrad et al., "Sex Differences and Similarities in Job Attribute Preferences: A Meta-analysis," *Psychological Bulletin* 126 (2000): 593–641.

5. National Bureau of Economic Research, Working Paper #5188, C. Goldin, *Career and Family: College Women Look to the Past* (Cambridge, Mass.: National Bureau of Economic Research, 1995).

6. J. T. Lindley, M. Fish, and J. Jackson, "Gender Differences in Salaries: An Application to Academe," *Southern Economic Journal* 59 (1992): 241–59.

7. K. C. Browne, *Divided Labours: An Evolutionary View of Women at Work* (New Haven, Conn.: Yale University Press, 1999).

8. S. A. Metzley Davies, "Women Above the Glass Ceiling: Perceptions on Corporate Mobility and Strategies for Success," *Gender and Society* 12 (1998): 339–55.

9. Department of Commerce, *Statistical Abstract of the United States* (Washington, D.C.: GPO, 1998).

10. C. P. Benbow, "Sex Differences in Mathematical Reasoning Ability in Intellectually Talented Preadolescents: Their Nature, Effects, and Possible Causes," *Behavioral and Brain Sciences* 11 (1988): 169–232.

11. D. Kimura, *Sex and Cognition* (Cambridge: MIT Press, 1999).

12. S. S. Brehm and S. M. Kassin, *Social Psychology* (Boston: Houghton Mifflin, 1990).

13. B. B. Whiting and J. W. M. Whiting, *Children of Six Cultures* (Cambridge: Harvard University Press, 1975).

14. D. C. Geary, *Male, Female: The Evolution of Human Sex Differences* (Washington, D.C.: American Psychological Association, 1998).

15. A. Shelton and D. John, "The Division of Household Labor," *Annual Review of Sociology* 22 (1996): 299–322.

16. Geary, *Male, Female.*

17. A. McMahon, "Blokus Domesticus: The Sensitive New Age Guy in Australia," *Journal of Australian Studies* (March 1998): 147–57.

18. T. N. Greenstein, "Husband's Participation in Domestic Labor: Interactive Effects of Wives' and Husbands' Gender Ideologies," *Journal of Marriage and the Family* 58 (1996): 585–95.

19. Geary, *Male, Female.*

20. J. Kalat, *Biological Psychology*, 5th ed. (Pacific Grove, Calif.: Brooks/Cole, 1995).

21. Geary, *Male, Female.*

22. Ibid.

23. Ibid.

CHAPTER NINE: THE MARRIAGE MARKET AND SINGLE PARENTHOOD

1. N. Barber, *Why Parents Matter: Parental Investment and Child Outcomes* (Westport, Conn.: Bergin and Garvey, 2000).

2. Ibid.

3. S. J. South, "Racial and Ethnic Differences in the Desire to Marry," *Journal of Marriage and the Family* 55 (1993): 357–70.

4. Department of Health and Human Services, *Vital Statistics of the United States,* vol. 3, *Marriage and Divorce* (Washington, D.C.: GPO, 1988).

5. K. M. Lloyd and S. J. South, "Contextual Influences on Young Men's Transition to First Marriage," *Social Forces* 74 (1996): 1097–1119.

6. J. R. Flynn, *Asian Americans: Achievement Beyond IQ* (Hillsdale, N.J.: Lawrence Erlbaum, 1991).

7. Barber, *Why Parents Matter.*

8. Ibid.

9. National Bureau of Economic Research Working Paper #5188, C. Goldin, *Career and Family: College Women Look to the Past* (Cambridge, Mass.: National Bureau of Economic Research, 1995).

10. N. Barber, "The Role of Reproductive Strategies in Academic Attainment," *Sex Roles* 38 (1998): 313–23.

11. S. G. Corona and R. Tidwell, "Differences Between Adolescent Mothers and Nonmothers: An Interview Study," *Adolescence* 34 (1999): 91–97.

12. J. S. Musick, *Young, Poor, and Pregnant: The Psychology of Teenage Motherhood* (New Haven, Conn.: Yale University Press, 1993).

13. N. Barber, *Parenting: Roles, Styles, and Outcomes* (Commack, N.Y.: Nova Science, 1998).

14. Corona and Tidwell, "Differences Between Adolescent Mothers and Nonmothers."

15. Barber, *The Role of Reproductive Strategies.*

16. F. F. Furstenberg Jr., J. Brooks-Gunn, and S. P. Morgan, *Adolescent Mothers in Later Life* (New York: Cambridge University Press, 1998).

17. H. P. Samuels, "The Relationships Among Selected Demographics and Conventional and Unconventional Sexual Behaviors Among Black and White Heterosexual Men," *Journal of Sex Research* 34 (1997): 85–92.

18. Corona and Tidwell, "Differences Between Adolescent Mothers and Nonmothers."

19. N. Barber, "On the Relationship Between Country Sex Ratios and Teen Pregnancy Rates," *Cross-Cultural Research* 34 (2000): 26–37.

20. N. Barber, "Predicting Social Problems from Parental Investment Indicators at Birth Using U.S. Population Data" (unpublished paper).

21. M. A. Fossett and K. J. Kiecolt, "A Methodological Review of the Sex Ratio: Alternatives for Comparative Research," *Journal of Marriage and the Family* 53 (1991): 941–57.

22. N. Barber, "Marital Opportunity, Parental Investment, and Teen Pregnancy Rates of Blacks and Whites in U.S. States," *Cross-Cultural Research* 35 (2001): 263–79.

23. E. Anderson, *Streetwise* (Chicago: University of Chicago Press, 1990).

24. S. J. Ventura, S. C. Curtis, and T. J. Mathews, *Teenage Births in the United States: National and State Trends 1990–1996* (Hyattsville, Md.: National Center for Health Statistics, 1998).

25. Population Reference Bureau, *World Population Data Sheet* (Washington, D.C.: Population Reference Bureau, 1998).

26. Anderson, *Streetwise*.

27. T. Sowell, *Race and Culture: A World View* (New York: Basic Books, 1995). D. D'Souza, *The End of Racism: Principles for a Multiracial Society* (New York: Free Press, 1995).

28. Anderson, *Streetwise*.

29. Ventura et al., *Teenage Births in the United States*.

CHAPTER TEN: FASHION TRENDS AND MARRIAGE: ANOREXIA, BEARDS, AND SKIRT LENGTH

1. K. Hubbard, "Slip of a Girl: Princess Victoria, Heir to Sweden's Throne, Falls Victim to Anorexia," *People* 49, no. 1, 1998, p. 135.

2. Ibid.

3. A. Forsyth, *A Natural History of Sex* (New York: Charles Scribner, 1986). Also, C. P. McCormack, ed., *Ethnography of Fertility and Birth*, 2d ed. (Prospect Heights, Ill.: Waveland Press, 1994).

4. Ibid.

5. J. L. Anderson, "Was the Duchess of Windsor Right? A Cross-Cultural Review of the Socioecology of Ideals of Female Body Shape," *Ethology and Sociobiology* 13 (1992): 197–227.

6. N. Barber, "The Evolutionary Psychology of Physical Attractiveness: Sexual Selection and Human Morphology," *Ethology and Sociobiology* 16 (1995): 395–424.

7. B. Silverstein, "Possible Causes of the Thin Standard of Bodily Attractiveness for Women," *International Journal of Eating Disorders* 5 (1986): 907–16.

8. N. Barber, "The Slender Ideal and Eating Disorders: An Interdisciplinary 'Telescope' Model," *International Journal of Eating Disorders* 23 (1998): 295–307.

9. B. Barber, "Secular Changes in Standards of Bodily Attractiveness in Women: Tests of a Reproductive Model," *International Journal of Eating Disorders* 23 (1998): 449–54.

10. Silverstein et al., "Possible Causes of the Thin Standard."

11. C. L. Kleinke and R. A. Staneski, "First Impressions of Female Bust Size," *Journal of Social Psychology* 110 (1980): 123–34.

12. N. Barber, "Reproductive and Occupational Stereotypes of Bodily Curvaceousness and Weight," *Journal of Social Psychology* 139 (1999): 247–49.

13. Silverstein et al., "Possible Causes of the Thin Standard."

14. Barber, "The Slender Ideal and Eating Disorders." Also, Barber, "Secular Changes in Standards of Bodily Attractiveness."

15. D. Singh, "Adaptive Significance of Female Physical Attractiveness: Role of Waist-to-Hip Ratio," *Journal of Personality and Social Psychology* 65 (1993): 293–307.

16. N. Barber, "Secular Changes in Standards of Bodily Attractiveness in American Women: Different Masculine and Feminine Ideals," *Journal of Psychology* 132 (1998): 87–94.

17. Barber, "Reproductive and Occupational Stereotypes."

18. P. N. Stearns, *Fat History* (New York: New York University Press, 1997).

19. Barber, "Secular Changes in Standards of Bodily Attractiveness." Also, Barber, "The Slender Ideal and Eating Disorders."

20. D. M. Buss, *The Evolution of Desire: Strategies of Human Mating* (New York: Basic Books, 1994).

21. M. A. Mabry, "The Relationship between Fluctuations in Hemlines and Stock Market Averages from 1921–1971" (master's thesis, University of Tennessee at Knoxville, 1971).

22. N. Barber, "Women's Dress Fashion as a Function of Reproductive Strategy," *Sex Roles* 40 (1999): 459–71.

23. Buss, *The Evolution of Desire.*

24. K. Grammar, "The Human Mating Game: The Battle of the Sexes and the War of Signals" (paper presented to the Human Behavior and Evolution Society, Northwestern University, Evanston, Ill., June 1996).

25. D. Popenoe, *Disturbing the Nest: Family Change and Decline in Modern Societies* (Hawthorne, N.Y.: Aldine De Gruyter, 1988).

26. D. Buss, "Sex Differences in Human Mate Preferences: Evolu-

tionary Hypotheses Tested in 37 Cultures," *Behavioral and Brain Sciences* 12 (1989): 1–49. Also, R. A. Posner, *Sex and Reason* (Cambridge: Harvard University Press, 1992).

27. Buss, *The Evolution of Desire*.

28. R. D. Guthrie, *Body Hot Spots* (New York: Van Nostrand Reinhold, 1976).

29. R. J. Pellegrini, "Impressions of the Male Personality as a Function of Beardedness," *Psychology* 10 (1973): 19–33.

30. N. Barber, "Mustache Fashion Covaries with a Weak Marriage Market for Women," *Journal of Nonverbal Behavior* (in press).

31. Ibid.

32. A. Hellstrom and J. Tekle, "Person Perception through Facial Photographs: Effects of Glasses, Hair, and Beard, on Judgments of Occupation and Personal Qualities," *European Journal of Social Psychology* 24 (1994): 693–705. Also, J. Klapprott, *"Barba facit magistrum*: An Investigation into the Effect of a Bearded University Teacher on His Students," *Psychologie Schweizeriscche Zeitschrift fuer Psychologie und ihre Abwendungen* 35 (1976): 16–27; S. Roll and J. S. Verinis, "Stereotypes of Scalp and Facial Hair as Measured by the Semantic Differential," *Psychological Reports* 28 (1971): 975–80.

Index